W0081755

Objects in Motion
GLOBALIZING TECHNOLOGY

Edited by Nina Möllers and Bryan Dewalt

artefacts
STUDIES IN THE HISTORY OF
SCIENCE AND TECHNOLOGY

volume 10

Managing Editor
Martin Collins, Smithsonian Institution

Series Editors
Robert Bud, Science Museum, London
Bernard Finn, Smithsonian Institution
Helmuth Trischler, Deutsches Museum

Smithsonian Institution
Scholarly Press

Washington, D.C.
2016

The series "Artefacts: Studies in the History of Science and Technology" was established in 1996 under joint sponsorship by the Deutsches Museum (Munich), the Science Museum (London), and the Smithsonian Institution (Washington, DC). Subsequent sponsoring museums include Canada Science and Technology Museum; Istituto e Museo Nazionale di Storia della Scienza; Medicinsk Museion Kobenhavns Universitet; MIT Museum; Musée des Arts et Métiers; Museum Boerhaave; Národní Technické Museum, Prague; National Museums of Scotland; Norsk Teknisk Museum; Országos Mıszaki Múzeum Tanulmánytára (Hungarian Museum for S&T); Technisches Museum Wien; Tekniska Museet–Stockholm; The Bakken; Whipple Museum of the History of Science.

Published by
SMITHSONIAN INSTITUTION SCHOLARLY PRESS
P.O. Box 37012, MRC 957
Washington, D.C. 20013-7012
www.scholarlypress.si.edu

Copyright © 2016 by Smithsonian Institution

Cover image: Interior view of the Centennial Hall in Wrocław, Poland. Deutsches Museum Munich, archives, CD_64744.

Library of Congress Cataloging-in-Publication Data:
Names: Möllers, Nina, editor. | Dewalt, Bryan, 1959- editor.
Title: Objects in motion : globalizing technology / edited by Nina Möllers and Bryan Dewalt ; managing editor, Martin Collins, Smithsonian Institution.
Description: Washington, D.C. : Smithsonian Institution Scholarly Press, 2016. | Series: Artefacts : studies in the history of science and technology ; volume 10 | Includes bibliographical references.
Identifiers: LCCN 2015046813 | ISBN 9781935623977 (paperback)
Subjects: LCSH: Technology—Social aspects. | Technology transfer | Communication and traffic. | Globalization. | BISAC: TECHNOLOGY & ENGINEERING / History. | SCIENCE / History. | HISTORY / Essays.
Classification: LCC T14.5 .O24 2016 | DDC 303.48/3-dc23
LC record available at http://lccn.loc.gov/2015046813

ISBN: 978-1-935623-97-7

Printed in the United States of America

⊗ The paper used in this publication meets the minimum requirements of the American National Standard for Permanence of Paper for Printed Library Materials Z39.48–1992.

Contents

Technology Transfers: East–West, North–South

Dimensions of Globalization: Objects and Identities

Series Preface

Science and technology have been defining elements of the modern era. They have entered into our lives in large and small ways—through broad understandings of the universe and in the tools and objects that make up the texture of everyday life. They have been preeminent activities for organizing expertise and specialized knowledge, in defining power and progress, and in shaping the development of nations and our relations with others across the planet. In 1996, the Smithsonian Institution, the London Science Museum, and the Deutsches Museum formed Artefacts to emphasize the distinctive role that museums—through their collection, display, and, especially, study of objects—can play in understanding this rich and significant history.

Artefacts has two primary aims: to take seriously the material aspects of science and technology through understanding the creation and use of objects historically and to link this research agenda to the exhibition and educational activities of the world community of museums concerned with the intimate connections among material culture, the history of science and technology, and the transnational. The effort gradually has gained footing: Artefacts holds an annual conference and has expanded its formal organization to include fourteen cosponsors (listed on this volume's copyright page). This expanded community, composed primarily of European and North American museums, provides opportunity for more robust professional conversation and broadens the range of local and national historical experiences of science and technology represented in Artefacts. Not least, Artefacts has created a fruitful interplay between scholarly research and museum practice. Aided by its Advisory Editorial Board, it publishes this book series, which, in conjunction with annual meetings, has helped stimulate a broader turn toward material-based research in scholarship and its use in museum collecting and exhibitions.

The Artefacts community believes that historical objects of science and technology can and should play a major role in helping the public understand science and technology: the ingenuity associated with these activities, their conceptual underpinnings, their social roles, and their local and global connotations. We welcome other museums and academic partners to join our effort.

Martin Collins
National Air and Space Museum, Smithsonian Institution
Series Managing Editor

Introduction

In the twenty-first century, technology has crossed borders and become global. It no longer surprises us to see Africans typing away at Chinese computers or Arabs using American mobile phones to send videos of protesters to all corners of the earth. Objects may be engineered in America, built in Asia, used in Australia, and eventually disposed of in Africa. Whether people, products, or knowledge, everything seems to be on the move, and almost any phenomenon that we encounter can be supplemented with the adjective "global": from global communication and consumer culture(s) to global trade, tourism, and terrorism to global pollution and global warming. Technology in its manifold shapes and manifestations plays a part in most of them, whether as cause, medium, or solution.

The movement of technology is first and foremost about spatial transfer. In the nineteenth and twentieth centuries, technology moved largely along the axes of east–west and north–south. Today, the picture has become more complicated: technological objects and, along with them, people, knowledge, and entire companies are moving in multidirectional ways. Although well-established paths of technology transfer are still active today, in many cases globalization has produced circular rather than linear movements of objects, people, and ideas. Objects are no longer just transferred from one region or nation to another: their local adaptations have repercussions on the point of origin as well.

Despite recent developments in globalization, wandering technological objects are by no means a phenomenon of the twenty-first century. In fact, the history of technology knows uncountable instances where objects, people, and knowledge moved across territorial borders. Technology transfer, be it drilling, spinning machines, or electron microscopy, has been a well-researched topic for generations of historians.[1] Similarly, studies of diverse forms of appropriation of a particular technology in different social and cultural contexts abound, as, for example, in the case of the cellular phone.[2] Although the objects that now move around the globe may have changed, they remain central actors in the ongoing dynamics of globalization. By their double nature as both material entity and symbol, they produce, reproduce, and react to globalizing forces.

As we increasingly understand artifacts not just for their materiality but for their meanings, our explorations take us beyond modes of functioning into the social and cultural realms of technology. Although the "brute intransigence of matter" suggests a universality conducive to a globalized world, the concomitant "plasticity of meaning, bound to specific times and places," makes the appropriation of artifacts particular and local and thus their study as globalized objects challenging.[3] Nevertheless, in a world of globalized production and consumption cycles, we must take a closer look at both the ways technological objects have been engineered, built, and sold and how they have been perceived and appropriated by global audiences.

Globalization also challenges our view of nation-states and the objects we associate with them. As Bruce Mazlish has argued, modern globalization owes its origins to the emergence of nation-states, who succeeded in building an international system in which global relations over trade, territory, and power were contested and resolved. Yet the global forces that this international system unleashed have increasingly subverted the power of the nation-states themselves.[4] As trade barriers have fallen and the reach of multinational corporations has increased, technological artifacts are increasingly perceived as international, transnational, global objects. Although products of national politics, science, and economics, technological objects now oscillate between national and global points of reference. Objects once identified with their countries of origin have acquired new meanings in their global travels.

An important theme explored by the essays in this volume is the question of how technology has been received and in what ways it has been "localized" in the process of adaptation. Moving around in a global cultural economy, objects are taken out of their territorialized and localized frames of meaning, only to be re-embedded into and thus resignified by new contexts. In exploring this question it is helpful to apply Arjun Appadurai's concept of a global cultural economy made up of interconnected and interdependent "landscapes": the ethnoscape, financescape, technoscape, mediascape, and ideoscape. Of special interest to us is the technoscape, a fluid "global configuration . . . of technology both high and low, both mechanical and informational, [that] now moves at high speeds across various kinds of previously impervious boundaries."[5] This technoscape acts on and is in turn constructed by the ethnic, economic, cultural, and ideological dimensions of globalization. This interplay of landscapes, however, is not linear but disjunctive, the configuration of elements varying from place to place so that the experience of a tabla maker in Banaras is distinct from, yet somehow related to, that of a global tractor manufacturer in Mannheim. Although the border-crossing quality of global cultural flows thus suggests disconnection from spatial and temporal anchors, it is rather the opposite. Appadurai's landscapes are "inflected very much by the historical, linguistic and political situatedness of different sorts of actors" and "navigated by agents who both experience and constitute larger formations."[6] We need to focus on these diverse actors in their locally and historically specific circumstances, be they particular national states, multinational corporations, local craftspeople, individual users, or the objects themselves.

Applying Appadurai's macroethnographical approach to the history of technology, two premises emerge for the study of globalized and globalizing technological artifacts. First, the globalization of technology involves the movement of both objects and of ideas, and objects are to be taken seriously in the duality of their materiality and their meaning. Second, objects change and are being changed while they actively engage in globalization. The following papers explore in varying degrees this pairing of the material and the symbolic and take objects as departure point, reference point, and centering anchor for the analysis of the complex interacting processes of globalization and localization. Yet the variety of sources deployed by the authors (e.g., material culture, oral history, written texts, images) points to the fact that our understanding of an object is both contingent and contextual. Meanings are constructed with reference not just to the material thing but to fluid portraits of the economic, political, and cultural landscapes in which it exists.

The first five essays in this volume explore the dynamics of technology transfer and global trade in the twentieth century. These essays deal with some of the foundational technologies of the modern era: concrete construction, scientific agriculture, mass manufacturing, motorized transportation, mining, and electronics. As such, they may be perceived as inhabiting Appadurai's technoscape in an uncomplicated way. Yet matters of ideology, economics, and symbolism point to broader and more complex movements that cannot be encapsulated by the simple, linear notion of "transfer."

Knut Stegmann's study of the adoption of American reinforced concrete construction techniques by a German firm in the early twentieth century uncovers a transatlantic movement not of objects or devices but of practice, namely, methods of organizing a complex building site. This movement of technique depended not on the formal exchange of knowledge through publications and patents but on personal observation by structural engineers working for a construction company. This in turn was facilitated by the increasing availability of relatively rapid, long-distance personal transportation and the support of a large, internationally active firm. At the same time, transfer was conditioned by local building practices, the skills of workers, and local building codes.

Johanna Conterio explores another example of east–west technology transfer, this time one that failed to occur. Unlike Stegmann's concrete builders, interwar Soviet agricultural scientists became aware of American orchard heaters largely through texts and a few exemplary objects. Their interest in this technology is set against a global diffusion and expansion in citrus growing and Stalin's drive for self-sufficiency in food. In this context, the reception of orchard heating was ambivalent. The simple American oil-burning buckets were effective but costly. Although transfer in this case was constrained by economics, Conterio also suggests that the heaters posed a challenge to the ascendant Soviet ideology of advance through the fundamental remaking of citrus fruits themselves.

This notion that technologies and objects bear potential symbolic importance is taken up by Thomas Schuetz in his study of the transfer of wastewater treatment technology from western Germany to the Soviet Union. In this case two layers of symbolic exchange underlaid the physical transfer of the huge steel reaction vessels. The first was the Soviet regime's attempt to transplant one of global capitalism's iconic, charismatic technologies of the mid-twentieth century: the automobile. The second was the adoption, seemingly without modification, of the entire production system to make the vehicle, right down to the plant for recovering and treating wastewater. Also of interest is Schuetz's focus on a medium-sized business, operating in an environment of intense local competition, as an agent for cross-border technology transfer.

David McGee and Rian Manson are also interested in the dynamics of technology transfer in the post-1945 world and provide a convincing case for the role of the Cold War as an impetus for globalization. Rather than movement along an east–west axis, McGee and Manson explore movement from north to south, in the context of decolonization in India and western responses to the perceived threat of communist expansion. In this context, the export of obsolescent steam locomotives from Canada to India took on great symbolic importance. The huge objects materialized and stabilized a new network of relations between two countries that heretofore had had no relations at all. In the process, concepts of "development" and aid to "developing countries" took root.

Matthew Hockenberry is also interested in evolving global relations in the transition from imperialism to decolonization within the context of global capitalism. Tin has been a globally traded, fundamental material of civilization since the Bronze Age. Yet in the twenty-first century, as the main component in the solder that binds electronic circuits, tin now is indispensable to the infrastructure of globalization. Its extraction from the "tin islands" of Indonesia dates from the period of Dutch imperialism and continues under nominal control of the postcolonial state. Yet production of this global industrial staple continues to rest to a great extent on hand labor, unlicensed resource exploitation, and local, noncorporate forms of organization.

The second set of essays explores the fluid and sometimes fraught dynamics of identity where the local, the national, and the global collide. Technological objects, their design, and the meaning assigned to them are often central to these dynamics, thus neatly demonstrating again the variable and locally particular ways that the technical, economic, political, and cultural landscapes of globalization intersect.

Allen Roda's case study of a type of drum made in the north Indian town of Banaras is, on one level, an exploration of the explicitly cultural dimensions of globalization. The distinct percussive tones of the tabla, once identified with Hindustani classical music, are now at the beating heart of transnational musical communities. What is more fascinating, however, is how tabla themselves have followed the music on its global journey. In the process, the meaning, value, and even physical form of the drums have changed. Among other things, this has disrupted relationships between makers and users, and drums that were once singular objects arising from interpersonal relations have become commodities whose very materiality is transformed to meet the exigencies of global markets.

National identity figures more explicitly in Harun Kaygan's study of automatic Turkish coffee makers. The everyday act of making coffee can become infused with questions of cultural authenticity. Contrary to prevalent theories that ascribe to globalization a homogenizing or hybridizing impact on local and national cultures, Kaygan argues that cultural practices strongly identified with a nation can be enlisted in commercial technological projects aimed at the global market. In the case of Turkish coffee, the technique of making the distinctive brew has been "black boxed" in an automatic process that can be commoditized for export in the form of a machine.

Bryan Dewalt's study of a singular object, a gift made from a recycled industrial drum, uncovers a fascinating microstory of globalization. To a large extent, this is a story of identities. On the one hand, national identity can be constructed as a conscious form of resistance against globalizing and localizing forces. This process of construction can include the selection of particular objects and technologies, in this case the canoe, to represent the idea of the nation. Yet these same icons can be appropriated and redeployed for different purposes, including the creation of transnational communities. Dewalt also reminds us that globalization plays out at the level of personal relationships, and these can be materialized in the exchange of objects.

The question of national technological identity also figures in Oliver Schmidt's study of farm machinery. Heinrich Lanz, the German manufacturer of the Bulldog line of tractors, self-consciously positioned its machines as global products, at home no less in the tropics than on the steppes. Yet this global image was mainly for domestic consumption, feeding a German self-image of technological prowess that was useful primarily for sales at home. Ironically, Heinrich Lanz eventually would become an actor in a global wave of consolidation among farm machinery manufacturers that would transform both its product line and its identity.

The final contribution to this volume addresses the role of museums as actors in the dynamics of globalization. As repositories of objects and as locations where their meanings are explored, museums must respond to the challenge that globalization poses for institutions generally oriented to local and national points of reference. The Internet and World Wide Web have been among the most important vehicles of globalization in recent decades, but museums have often struggled to effectively exploit them for their own purposes. Kimberly Coulter's account of the Inventing Europe project shows how several museums are harnessing digital technology for a collaborative, multinational exploration of the technological processes and narratives of European integration. In the process, Inventing Europe demonstrates how objects with very particular life histories can be redeployed in multiple narratives that transcend the local.

Museums and human-made objects meet in the study and interpretation of material culture, and technological objects are an essential part of this record of cultural construction preserved by museums. It is hoped that this volume, by bringing together the methods and objects of study from anthropology and the history of technology, will contribute to a broader and deeper engagement by museum scholars with the multifaceted dynamics of globalization in the past and the present. In doing so we also hope that this volume will help to integrate the study of technological change and the discourse around cultural change and to ground both in the common materiality and contingent meanings of objects.

Bryan Dewalt
Director of Curatorial Division, Canada Science and Technology Museums Corporation

Nina Möllers
Special Exhibition Project Curator, Deutsches Museum

Notes

1. William J. Pike, "Drilling Technology Transfer between North America and the North Sea: The Semi-submersible Drilling Unit," *History of Technology* 16 (1994): 1–14; Karen Johnson, "Innovation and Technology Transfer during the Cold War: The Case of an Open-End Spinning Machine from Communist Czechoslovakia," *Technology and Culture* 48, no. 2 (2007): 249–285; Nicolas

Rasmussen, "What Moves When Technologies Migrate? 'Software' and Hardware in the Transfer of Biological Electron Microscopy to Postwar Australia," *Technology and Culture* 40, no. 1 (1999): 47–73.

2. Anandam Kavoori and Noah Arceneaux, eds., *The Cell Phone Reader: Essays in Social Transformation* (New York: Lang, 2006); Heike Weber, *Das Versprechen mobiler Freiheit: Zur Kultur- und Technikgeschichte von Kofferradio, Walkman und Handy* (Bielefeld: transcript, 2008); Akiba A. Cohen, *The Wonder Phone in the Land of Miracles: Mobile Telephony in Israel* (Cresskill, NJ: Hampton Press, 2008); Raul Pertierra, *Transforming Technologies, Altered Selves: Mobile Phone and Internet Use in the Philippines* (Manila: De La Salle University Press, 2006); Russell Southwood, *Less Walk, More Talk: How Celtel and the Mobile Phone Changed Africa* (Hoboken, NJ: Wiley, 2009).

3. Lorraine Daston, "Speechless," introduction to *Things That Talk: Object Lessons from Art and Science*, ed. Lorraine Daston (New York: Zone Books, 2004), 16.

4. Bruce Mazlish, *The New Global History* (New York: Routledge, 2006), 9.

5. Arjun Appadurai, "Disjuncture and Difference in the Global Cultural Economy," *Public Culture* 2, no. 2 (1990): 8.

6. Appadurai, "Disjuncture and Difference," 7.

Bibliography

Appadurai, Arjun. "Disjuncture and Difference in the Global Cultural Economy." *Public Culture* 2, no. 2 (1990): 1–24.

Cohen, Akiba A. *The Wonder Phone in the Land of Miracles: Mobile Telephony in Israel.* Cresskill, NJ: Hampton Press, 2008.

Daston, Lorraine, ed. *Things That Talk: Object Lessons from Art and Science.* New York: Zone Books, 2004.

Johnson, Karen. "Innovation and Technology Transfer during the Cold War: The Case of an Open-End Spinning Machine from Communist Czechoslovakia." *Technology and Culture* 48, no. 2 (2007): 249–285.

Kavoori, Anandam, and Noah Arceneaux, eds. *The Cell Phone Reader: Essays in Social Transformation.* New York: Lang, 2006.

Mazlish, Bruce. *The New Global History.* New York: Routledge, 2006.

Pertierra, Raul. *Transforming Technologies, Altered Selves: Mobile Phone and Internet Use in the Philippines.* Manila: De La Salle University Press, 2006.

Pike, William J. "Drilling Technology Transfer between North America and the North Sea: The Semi-submersible Drilling Unit." *History of Technology* 16 (1994): 1–14.

Rasmussen, Nicolas. "What Moves When Technologies Migrate? 'Software' and Hardware in the Transfer of Biological Electron Microscopy to Postwar Australia." *Technology and Culture* 40, no. 1 (1999): 47–73.

Southwood, Russell. *Less Walk, More Talk: How Celtel and the Mobile Phone Changed Africa.* Hoboken, NJ: Wiley, 2009.

Weber, Heike. *Das Versprechen mobiler Freiheit: Zur Kultur- und Technikgeschichte von Kofferradio, Walkman und Handy.* Bielefeld: transcript, 2008.

Technology Transfers:
East–West, North–South

Globalizing Building Technique
The Centennial Hall in Wrocław

Knut Stegmann
Research Fellow

*LWL-Denkmalpflege,
Landschafts- und Baukultur in
Westfalen
Münster, Germany*

Concrete Construction as a Transnational Technological Development of the Nineteenth Century

The introduction and widespread use of concrete construction is one of the most significant technological developments of the nineteenth century, which was from the start characterized by transnational knowledge-building processes. It had its origins mainly in Europe, the precursor being experimentation with new cements, particularly in England and France from the end of the eighteenth century. These experiments resulted in the development of new mortars and, in particular, at the beginning of the nineteenth century of prefabricated parts like pipes and artistic products such as ornaments and figures (Figure 1).[1] For the dissemination of this (partly tacit) knowledge, personal exchange played a major role. Wolfgang König stated this fact for the transfer of technology in general at this time: "Written information, such as scientific publications, patents and licenses, generally only supplemented the exchange taking place from person to person."[2] An important element in the exchange of knowledge was study trips, the majority of which were made by the founders of cement and concrete companies or

AUSTIN & SEELEY,

NEW ROAD, LONDON,

(Corner of Cleveland Street.)

AUSTIN and SEELEY respectfully invite the attention of Builders, Masons, and others to their extensive Collection of Ornaments, manufactured in Artificial Stone, of their own peculiar Composition, without either the use of Roman Cement or the application of Heat. They are also ready to execute New Models on the lowest remunerating terms. Their present Stock consists of —

Capitals and Fluted Columns; Trusses, Brackets, and Modillions; the Royal Arms and Prince of Wales's Feathers; Centre Ornaments for Entablatures and Bas-Relievos; Balustrading and Coping, for which, as their work is waterproof, it is well suited; Rustic and Rough Stone Facing, and Pier Ornaments, such as Pine-Apples, &c.; Gothic Work in great variety, including Fonts, Communion Tables, and Screens; Tazzas and Vases, to the extent of nearly One Hundred Models; Flower-Boxes, and Garden-Border Edging; Fountains, from £6 and upwards; Monumental Urns; Figures—Statues from the antique, as well as some chaste subjects of modern·design, Animals, Birds, &c; Chimneys and Chimney-Pots, from 1 foot 10 inches to 10 feet high. (As these are so bulky, a portion of Roman Cement is introduced for economy's sake.)

N.B. A complete Specimen-Sheet of their Chimneys may be had by application to A. and S.

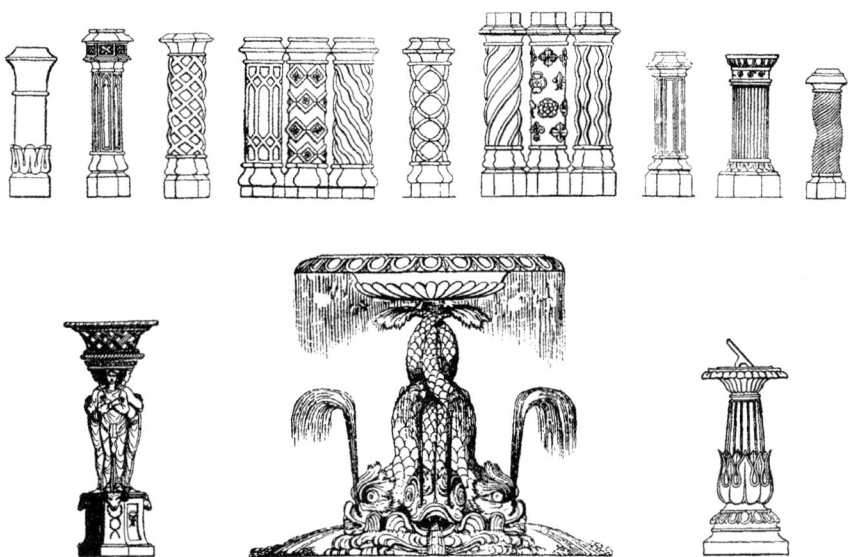

1.
Advertisement of the London-based company Austin and Seeley for its products in "artificial stone" in *The Builder* in 1842. *The Builder* 1, no. 1 (1842): 12.

on their behalf. The focus of many such trips was the English cement and concrete industry, which was already established in the first half of the nineteenth century and considered very advanced.[3] Other important destinations of these trips, particularly in the second half of the nineteenth century, were (world) exhibitions, which also facilitated an early transatlantic transfer of knowledge.[4] American experts who visited the exhibition stands of, for instance, French concrete pioneer François Coignet (1814–1888) at the 1867 Paris world exhibition in Paris reported in detail.[5]

Other important protagonists in the transnational knowledge exchange were the large concrete construction companies that had been established in the second half of the nineteenth century. Unlike the traditional tradesmen and small building contractors, these were soon working across national borders in Europe and thus spreading technological innovations.[6] Further important protagonists in this process were the sellers of what were known as construction "systems" (e.g., Monier System, Hennebique System). In particular, the French pioneers in reinforced concrete, Joseph Monier (1823–1906) and François Hennebique (1842–1921), ensured their ideas spread to countless countries by awarding licenses.[7] Even so, the development of concrete construction was not always based on linear adaptation processes and circulating knowledge. In early concrete construction many pioneers also worked simultaneously on the same questions at different places—often without knowing about each other or actually being in competition with each other.

Initially, the American construction industry not only imported European Portland cement in large quantities compared with the domestic cement production,[8] but it was also strongly based on European knowledge in material technology and construction in concrete.[9] This focus on European methods may also be connected to the initially low importance of the new construction material in the United States and the less challenging areas of application.[10] It was only at the end of the nineteenth century that domestic American cement production grew rapidly[11] and the large concrete construction companies with it, such as that of English-born Ernest L. Ransome (1852–1917).[12] These companies adapted concrete construction for the American market, with its high salaries and poorly qualified workers, by placing an emphasis on optimizing the building process.[13] Mechanization, standardization, prefabrication, and organizational improvements led to a gradual breakthrough of the new construction method in the United States in the early twentieth century. In Europe these innovative ideas were not widely adopted in large-scale concrete construction, apart from the discussion and sale of American machines around the turn of the century.[14] How and why the ensuing transatlantic exchange processes took place in the first decade of the twentieth century and how in this context European developments in material technology and static calculations came together with American developments in the building process are investigated below.

Case Study of the Centennial Hall in Wrocław 1910–1912

OVERVIEW OF THE CONSTRUCTION WORKS AND REQUIREMENTS

At least since its entry into the list of World Heritage Sites in 2006,[15] the Centennial Hall in Wrocław[16] has been accepted as an architectural masterpiece of the early twentieth century,

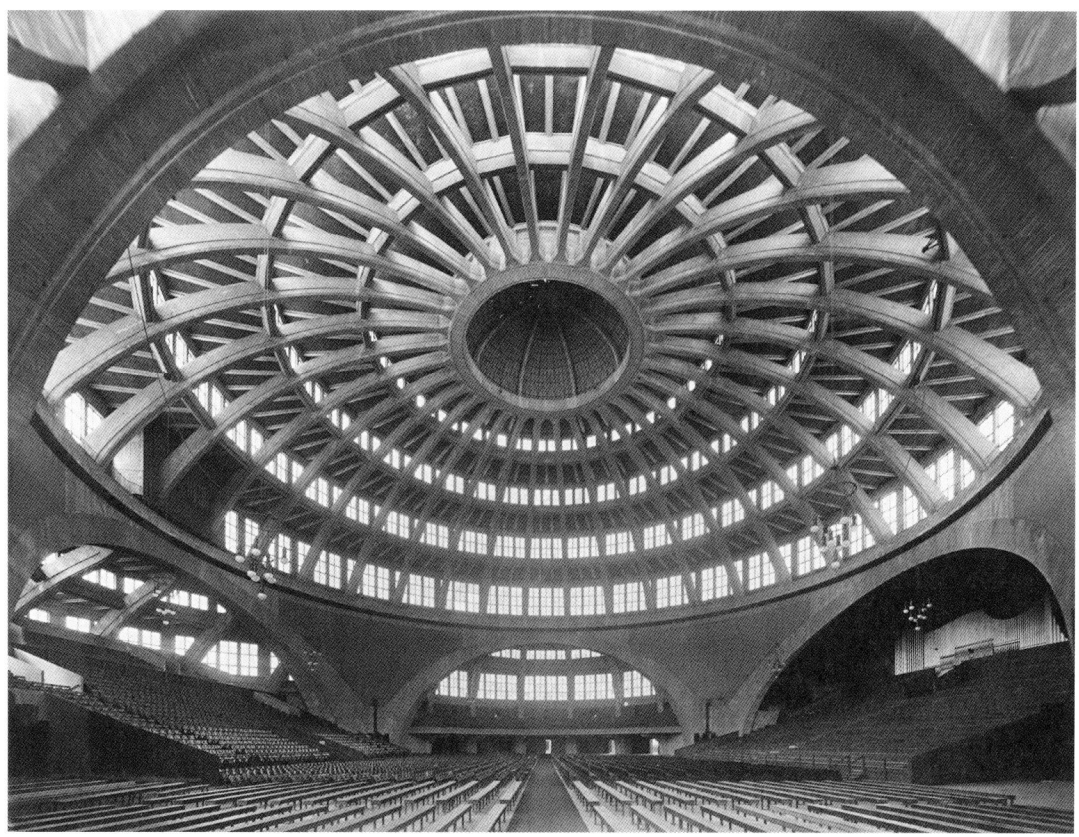

2.
Interior view of the Centennial Hall in Wrocław by Max Berg, erected by Dyckerhoff & Widmann and Lolat-Eisenbeton in 1911–1912, contemporary picture. Deutsches Museum Munich, archives, CD_64744.

largely due to the novel way in which the architect Max Berg (1870–1947) worked with the new building material of concrete in his design.[17] In the nineteenth century, concrete had been used almost exclusively for foundations and structural engineering. The Centennial Hall, constructed to mark the anniversary of the wars of liberation against Napoleon in 1913,[18] is one of the early large architectural representations completely and obviously constructed in concrete (Figure 2). Berg did more than use concrete as a hidden construction material. His meeting hall with several thousand seats took the sleek, reduced form of concrete construction out of industrial construction. Furthermore, it showed the material as exposed concrete with formwork marks.[19]

With a dome 65 meters in diameter, the Centennial Hall was far and away the largest of the concrete domes made by that time. This accomplishment meant it was a prestigious object for the concrete industry, which at the time was trying to break into the field of representational architecture.[20] However, the large project attracted critics, who doubted the stability of the concrete construction and criticized its high costs.[21] In order for reinforced concrete to compete at all against the long-established steel construction companies, the concrete construction companies had to optimize their construction processes in their tenders.[22] In this way they could ensure lower costs and also building completion in the brief space of time before the start of the centennial celebration. The tender for the structurally challenging central dome was finally won in August 1911 by the Dresden branch of the concrete construction company Dyckerhoff & Widmann. The Wrocław-based company Lolat-Eisenbeton was permitted to execute the hallway arranged around it.[23]

TRANSFER PROCESSES

Concrete building technique demonstrates many specific characteristics when compared with other (fabrication) technologies. These characteristics influenced its global transfers. First of all, concrete building technique in the German Reich primarily served the production of single products or perhaps small local series, which each required individual fabrication techniques (particularly during the time period examined here). Second, the customer dictated the place of production in most cases by choosing a construction site. Building technique had to take the specific local conditions of the site into account (e.g., climate, infrastructure, availability of building materials). At best, parts of the work, mainly construction materials and preproducts, could be prefabricated and stored at central locations.[24] Third, practical knowledge and experience also played a central role, given the heterogenic finished products. This hard to document knowledge was mainly transferred on the construction site rather than theoretically. Fourth, building technique is regulated more strictly than other technologies by local laws and authorities (building codes and site supervision). Finally, the long tradition of building technique means it is strongly shaped by (local) customs and building culture. These specific requirements, in particular the multifaceted local commitments, hindered the globalization of concrete construction. Thus, the Centennial Hall represents one of the first large concrete constructions ever in the German Reich that was influenced to a large extent by transatlantic transfer processes.[25] This distinction means it presents a good object for investigating early transfer processes that were starting points for the subsequent globalization of concrete building technique.

At the turn of the century, this transfer was still far removed from Arjun Appadurai's "technoscape," where technology "moves at high speeds across various kinds of previously impervious boundaries."[26] However, the example of the Centennial Hall demonstrates how globally active companies and faster means of transatlantic transportation leveled the playing field for early technological transfers in concrete building technique beyond the confines of Europe.[27] Neither in the published literature nor in the extensive estate of Max Berg[28] is there any indication of the input of the architect in the transfer processes. This finding is contrary to the widely purported impression that the designers of buildings largely initiated the exchange processes in concrete building technique.[29] As building and construction history focus largely on architects and engineers, the absence of these planners in the transfer processes may be one reason for the low interest of researchers in this part of the building's history so far.[30] The sources for the Centennial Hall suggest that it was rather the concrete construction company Dyckerhoff & Widmann that played an important role in the process. The importance of such companies as transnational actors has been, at least in regard to the German Reich, largely overlooked.[31] If we look at the history and structure of Dyckerhoff & Widmann, the important role the company played in the transfer processes is strongly indicated.

This concrete construction company was created in 1865 as a sort of offshoot of the cement factory Dyckerhoff & Söhne,[32] established in Amöneburg a few years earlier. The cement factory was already a global player a few years after it was founded, with worldwide business contacts. Business trips led members of the owning Dyckerhoff family to India, China, Japan, Canada, and the United States in the nineteenth century.[33] The concrete company Dyckerhoff & Widmann was influenced by the international positioning of the sister company. From the start, the founders used their international business contacts from the cement factory to turn the concrete construction company into a cross-border enterprise, albeit one initially restricted to central Europe. In this case, the term "cross-border company" does not refer to just the development of marketing in neighboring countries. It also means the inclusion of foreign state-of-the-art technology in research and development. Eugen Dyckerhoff (1844–1924), a partner of the concrete construction company since 1866, traveled to concrete construction sites and precast concrete plants, as well as to exhibitions in France and Austria. On his trips he studied new processing and production methods systematically.[34] Importantly, these extended study trips benefitted not only from the preexisting business contacts from Dyckerhoff & Söhne's cement trade but also from the shorter traveling times created by new means of transportation, in particular railways and steamboats. In summary, the exchange processes primarily took place on a personal level (which also included the intermittent allocation of workers).

The transfer processes remained largely confined to Europe in the nineteenth and early twentieth centuries for different reasons. First of all, many of the important early pioneers worked in Europe and could easily visit each other. Traveling to America required lots of time and financial investment at a point when the continent did not (yet) embody an interesting market for the emerging concrete industry in the German Reich. However, U.S. trips made by Dyckerhoff family members ensured that Dyckerhoff & Widmann had firsthand information about American developments in concrete building construction, which was important during later development in the early twentieth century. These insights and the increasing coverage in the German Reich regarding the fast and efficient construction of tall buildings in the United States (although initially they were primarily steel structures) may have awakened the company's interest in American construction. The strong competitive pressure from steel construction companies in the German Reich[35] also forced the concrete industry to make efforts toward rationalization of the structural engineering and construction processes.

In 1909, Dyckerhoff & Widmann sent two employees, Franz Widmann (1882–1915), a nephew of the company's early partner Gottlieb Widmann, and Willy Gehler (1876–1953), the technical director at the Dresden branch and later a professor in Dresden, on expensive research trips to the United States. At that time, America had moved closer to Europe, thanks to the new, faster means of transportation. Nonetheless, a transatlantic journey was hardly an option for a normal architect because of time and financial constraints.[36] As a result, architects of that time lacked firsthand experience. This lack in turn led to contemporary articles on building in America that did not discuss American developments in concrete construction techniques and processes in detail.[37] As a result, it is less surprising that the architects in the early transfer processes played a less prominent role than the companies.

Franz Widmann's journey is comparatively well documented by an illustrated album,[38] as well as a published lecture on concrete construction in the United States, held at the general assembly of the German Concrete Association (Deutscher Beton-Verein) in 1911.[39] Widmann received the opportunity for the yearlong trip in 1909, after completing his structural engineering studies at the Technical College of Dresden and following his successful incorporation in the Dresden branch of Dyckerhoff & Widmann in 1907.[40] The latter accomplishment meant he already had firsthand experience of evaluating and executing large concrete buildings. This background knowledge explained his ability to assess what he saw in a different way:

> This is why, more than their [the Americans'] theoretical achievements, we recognize and marvel at the special talent with which they quickly and effectively accomplish their works, using practical design, a simple working disposition and the generous application of simple auxiliary machines.[41]

The album and the lecture also reveal Widmann's focused and selective vision on the study trip. They contain practically no discussion of American calculation methods. Instead, Widmann does very precisely document building techniques at the construction sites visited (for example, in New York, Seattle, and St. Louis) in (schematic) sketches and photos.[42] Just as in the European exchange processes, the central location for the transfers remained the construction site, in other words, building practice. The theoretical works of American rationalization experts Frederick Winslow Taylor (1856–1915) and Frank Bunker Gilbreth (1868–1924) on concrete building technique[43] are mentioned neither by Widmann nor by Gehler. This close focus on practice, as well as the numerous local connections, can be seen as a central hindrance to the globalization of building technique. According to Widmann, the digressive relationship of the cost of machines to handwork in America compared with the German Reich and the varying availability and qualifications of workers also made adaptations necessary.[44]

American building technique interested Widmann, as the evaluation of his report and his lecture demonstrate in two closely connected aspects. He was concerned first with the technique itself and second with its inclusion in the building processes. Where technique is concerned, he particularly documented the transportation facilities on the construction site. Numerous sketches and photos show horizontal and vertical transportation facilities, such as elevators, band conveyors, and slewing and cable cranes (Figure 3). These technologies were not new in themselves, although cable cranes were hitherto not used as hoisting devices on construction sites in the German Reich.[45] According to Widmann's depiction, it was through the systematic interaction of these machines that they became effective in developing the mechanized, American construction

3.
Machines on American construction sites as documented in Franz Widmann's unpublished picture album, 1910. Deutsches Museum Munich, archives, CD_65787.

site. For this reason, he documented the site in its overall context. Thus, one sketch in the documentation shows the site of individual storage spaces and machine assembly on a large bridge construction site. The optimized transportation facilities and routes on the construction sites were, according to Widmann's observations, in harmony with coordinated project planning. In his lecture, he mentions "the exceedingly practical project management and construction facilities, even in smaller companies."[46] This "practical project management" includes avoiding complex constructions in favor of extensive standardization.

Apart from the transportation facilities, Widmann also documented in his album other "auxiliary machines" largely unknown in the German Reich. One example is pneumatic tools, which were driven by compressed air from central compressor stations.[47] The presence of transportation facilities also directly influenced the architecture, according to Widmann's observations. A greater number of larger prefabricated components, which could be positioned on the construction site with the help of cable cranes, were used in the United States than in Europe.[48]

Although Widmann described the machine facilities and prefabrication as examples worthy of adaptation, he criticized the manual quality of the implementation. Thus, he explicitly mentions in his lecture that America has not succeeded in "manufacturing attractive surfaces from concrete and the plastered and unplastered facades are generally cracked and spotted."[49]

Willy Gehler also came to similar conclusions on the basis of his study trip.[50] In his opinion, American building technique was not a model for the complicated calculations of demanding concrete constructions, such as those required for the Centennial Hall, or for manual implementation. Rather, he saw the adaptable model for the German Reich in the mechanization of building sites, for which the planning was also coordinated. He attributed the "record breaking achievements of the Americans in reducing construction times for large buildings mostly to the clear construction schedule and careful preparation."[51] Moreover, he stated, "the Americans are ahead of us in the generous facilities on their construction sites and in the configuration and handling of their hoisting devices."[52]

ADAPTATION PROCESSES

These study trips showed that American concrete construction could serve as an adaptable example in only a few areas, in particular the mechanization of construction sites. Further restrictions to transfers resulted from differing initial requirements, which made extensive adaptation processes necessary. The fact that Dyckerhoff & Widmann did try to adapt the American model in Wrocław is due to more than the chronological coincidence of the study trips mentioned. Rather, the Centennial Hall was predestined for such attempts for several reasons, predominantly economic. The adoption of American building technique first promised considerable savings potential for large-scale buildings. Second, the company hoped to achieve a faster completion time using the transferred technology, which would not only save on salaries but also promised additional profit. The city of Wrocław offered Dyckerhoff & Widmann a bonus of 300 marks for each day that the company finished ahead of schedule, whereas a delayed completion came with a penalty of 500 marks for each day over the deadline.[53] The saying "time is money" applied here not only in a figurative sense.[54] Third, but no less important, the Centennial Hall was a prominent, prestigious building, with which the company wanted to present itself as a high-performance and progressive firm that could compete with the "record-breaking American achievements."[55] Corresponding to the observations made on the American construction sites, new technology was also linked to process alterations during the transfer and adaptation processes for the Centennial Hall.

The most important innovation from the transfer processes was related to the transportation facilities. At the heart of these facilities was the construction of a cable crane to transport building materials vertically, just as Gehler and Widmann had seen on many large American construction sites. These machines had not been used in building construction in the German Reich because transporting heavy loads over workers' heads was not permitted.[56]

The use of cable cranes on the construction site therefore required two steps. The first was adjusting the local guidelines and codes, which was achieved by acquiring special permits for overhead transport,[57] as well as submitting detailed building applications for the crane construction, with detailed static analyses according to local building codes.[58] The second step was technical adaptation. Gehler took inspiration from the American construction site cable crane model here. German knowledge was also in abundance for cable crane technology; the Leipzig company Adolf Bleichert & Co. was one of the globally leading companies in cable cranes for goods transportation.[59] But Bleichert's cranes were hardly ever used on building sites. The equipment planned by Gehler consisted of two cable cranes with their steel cables pulled between a 52-m-high stationary central tower and two 14-m-high mobile towers (Figure 4). The lower towers could be moved in a circular direction around the ground plan of the Centennial Hall, which was laid with tracks. This design meant the construction site could be serviced with precision. However, our interest concerns less the technical characteristics of the site (which Gehler detailed in his publication on the project)[60] and more how the new technology was simultaneously included in the organizational structures of the construction site. Gehler's construction site ground plan shows that rail tracks horizontally connected the cable cranes with the field factories arranged around the hall's ground plan as supply facilities (Figure 4). The new transportation system enabled the use of larger, prefabricated parts on the basis of American models. In contrast to the standardized American (industrial) buildings, which could be put together entirely from prefabricated elements,[61] the Centennial Hall was a highly unique construction. Thus, the application of prefabrication was limited to delicate window elements for the stepped superstructure of the outer dome (Figure 5). These elements were not poured at great heights but prefabricated on the ground in a field factory in forming boxes and then put in position using the cable crane. Willy Gehler noted, with no small amount of pride, that "each small pillar took only 10 minutes to position."[62]

As a further example of transfer processes, we should note the use of pneumatic tools (in particular the air rammer), fed by a central compression system. Dyckerhoff & Widmann was also exploring new ground with this system because there was little experience with such tools in the German Reich at that time.[63] More concrete strength was achieved in some construction parts with pneumatic rammers. However—and this is also a pivotal point for the other transfers mentioned—the technical improvements were not the only decisive criteria. First and foremost, economic conditions were the trigger for the transfers. Increasing salaries in Germany, particularly in relation to the costs of using machines, promised decisive savings through mechanization.[64] Thus, pneumatic rammers replaced the time-consuming concrete compaction by hand with iron hand tampers.

The use of new machines not only changed building processes but also influenced the project planning for the Centennial Hall. Expensive holding time for machines was also supposed to be reduced through parallel work, as in the American model. Crucial here was dividing the hall into two construction sections: the base level and the dome. This division enabled the company

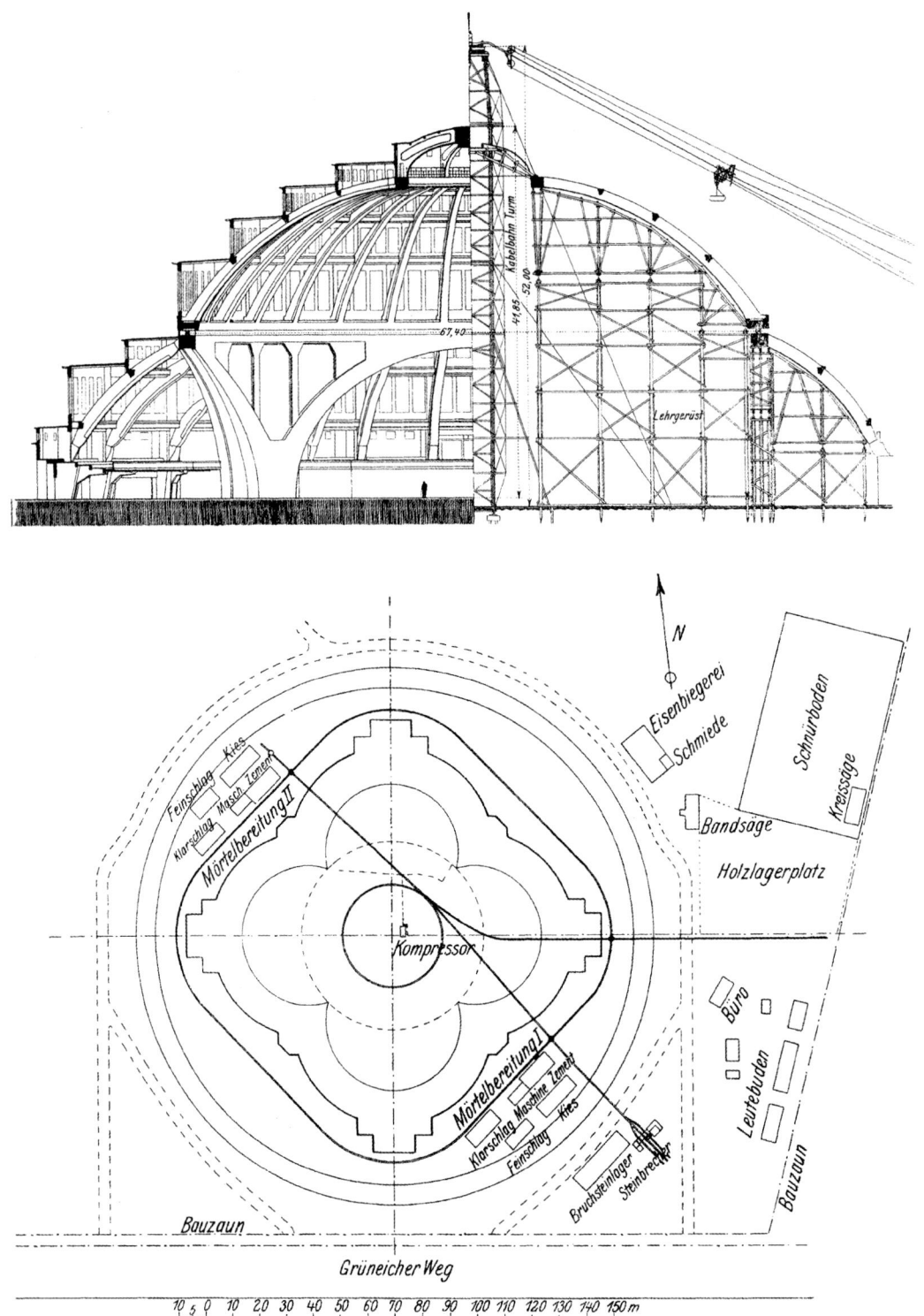

4.
Mechanization and organization of the construction site according to American models: construction site plan of the Centennial Hall in Wrocław by Willy Gehler. Ground plan and cross section. Günther Trauer and Willy Gehler, *Die Jahrhunderthalle in Breslau: Berechnung, Konstruktion und Bauausführung* (Berlin: Julius Springer, 1914), 59, 62a.

5.
Prefabricated elements put in position by the cable crane: window element for the stepped superstructure of the Centennial Hall in Wrocław, ca. 1911. Deutsches Museum Munich, archives, CD_65787.

to strip the forms of the base level before the dome was completed and then carry out the interior finishing of the base layer parallel to the construction of the dome. As Franz Widmann described in his lecture, this kind of parallel working was a preferred method in American high-rise construction to speed up the construction process.[65] Execution in two construction sections also meant the scaffolding and forming lumber could be reused, where a previously created plan ensured effective reuse.[66]

With the adaptation of American structural engineering methods, Dyckerhoff & Widmann was able to consign the building shell of the Centennial Hall in December 1912, six weeks ahead of the completion date agreed upon with the city of Wrocław.[67]

Contemporary Evaluation and Acceptance

Because of its political importance and also because of the unusual architecture and demanding construction, the Centennial Hall was extensively discussed in contemporary specialist publications. The company Dyckerhoff & Widmann itself directed the focus of the technical community on the hall's innovative building technique and the transatlantic transfer processes. Thus, its delegate, Willy Gehler, gave a lecture on that topic at the general assembly of the German Concrete

Association in 1913. The text of the lecture was published several times[68] and was thereafter quoted in many articles relating to the building technique of the Centennial Hall. It is conspicuous that Gehler repeatedly creates a demonstrative, rhetorical connection to American building technique with the prestigious Centennial Hall building.[69] This emphasis shows how dominant the concept of "progressive America"[70] was, although, for the reasons already outlined, the concrete construction companies were late in turning their attention to American developments in building technique in comparison to other industry sectors. Already in the 1870s, the machine industry, in particular Berlin industrialists Ludwig Loewe, Werner Siemens, and Emil Rathenau, had production sites set up according to American models, with Siemens using what was referred to as the "American hall" (*amerikanischer Saal*).[71]

For Gehler, the successful transfer processes lay less in technical improvements and more in economic ones. The innovations reduced the amount of manual work and thus the dependency of companies on specialist workers. In Gehler's opinion the decrease in manual work was necessary as "in concrete construction, as everywhere in the building industry, the trained, specialist workers are becoming increasingly rare and we, just like the Americans, must turn more and more to the use of machines."[72] At the same time, the technology transfer from America provided the basis for the increased working pace the companies wanted. Gehler describes this as being in no way a disadvantage for the workers, as did many other authors from the concrete and machine industry.[73] Workforce representatives, however, definitely took a critical stance toward the adaptation of American building technique to increase the pace of work. In *Grundstein*, the organ of the German construction workers' union, a contemporary article criticizes "the wild American way of building" and links it to accidents on construction sites.[74]

Despite this criticism, the following years saw a dramatic mechanization of large construction sites based on American models.[75] Dyckerhoff & Widmann was not the only company that used cable crane systems again and again in large constructions, as in the expansion of the Bahía Blanca naval base in Argentina (Figure 6).[76] By 1914, Hans Schäfer, an engineer from Darmstadt, referred to "the increased use of reaching cable cranes in modern times."[77] The compressed-air systems, the use of which Willy Gehler held in such high regard,[78] were used on more of the company's construction sites, as well as on those of other companies.[79] However, reference to technology transfers from America were quickly lost in discussions of German building technique, even when dealing with the example of the cable cranes at the Centennial Hall.[80] Instead, authors presented the innovations as achievements of the German machine construction industry, which can also be seen in the context of efforts to create a more pronounced national identity at the start of World War I.

Conclusion

The case study on the Centennial Hall throws the exchange processes between the German and American construction industries in the early twentieth century into new relief. The building can be seen as one of the starting points of globalized building technique in concrete construction.

6.
Building according to American models: cable cranes at the construction site of the expansion of the Bahía Blanca naval base. Project executed by Dyckerhoff & Widmann, 1911–1916. Staatsarchiv Leipzig, 20781, folder 446, picture 101.

The earlier modest transfers in this field had been restricted fundamentally to adaptations of European developments in material and construction technology in the United States. With the Centennial Hall, bidirectional exchange processes were established. Mainly, American building technique and organization of concrete construction sites were adapted in the German Reich. Economic considerations formed the backdrop for the transatlantic technology transfers. The heavily mechanized and optimized processes of American building technique promised important savings for the German concrete industry. These savings in turn made the industry more competitive in the face of increasing salaries and the high cost pressure from the competing steel construction companies.

The direct connection of technological transfers with economic questions explains a special characteristic of the transfer processes: not the architect, Max Berg, but the construction company, Dyckerhoff & Widmann, emerged as the central actor. The internationally active company sent employees on expensive study trips to the United States in advance because the specific characteristics of the building technique required firsthand experience.

In the Centennial Hall, the latest European findings in material science and static calculations of complex concrete constructions were linked with American building technique. Examples of the adaptation of the latter are the mechanized transportation facilities on the construction site (in particular the application of cable cranes), the use of new mechanical tools (pneumatic tools with a central compressor system), and partial prefabrication of building parts in field factories, which was conversely only possible with the transportation facilities offered by the construction site cable crane. However, the transfer was not limited to the technology itself. Its rational application by the adoption of American construction site organization formed another central factor in the adaptation processes. Overall, the example of the Centennial Hall shows the importance of large internationally active companies (the "construction industry") as actors during the early stages of globalized building technique in concrete construction.

Notes

This paper is based on the author's dissertation on the German concrete building company Dyckerhoff & Widmann. See Knut Stegmann, *Das Bauunternehmen Dyckerhoff & Widmann: Zu den Anfängen des Betonbaus in Deutschland 1865–1918* (Tübingen and Berlin: Wasmuth Verlag, 2014).

1. For an overview, see Gustav Haegermann, "Vom Caementum zum Zement," in *Vom Caementum zum Zement*, ed. Günter Huberti (Wiesbaden: Bauverlag, 1964), 36–53.

2. Wolfgang König, *Technikgeschichte: Eine Einführung in ihre Konzepte und Forschungsergebnisse* (Stuttgart: Franz Steiner Verlag, 2009), 158.

3. Study trips were already an important factor in the beginning of the German Portland cement industry. The chemist Hermann Bleibtreu (1824–1881) went on a study trip to England in 1853 before setting up the two earliest large Portland cement plants in the German states. His tricky industrial espionage is larger than life. See Bonner Portland-Zementwerke, *Ein Jahrhundert Bonner Zement: 1856–1956* (Bonn: Bonner Portland-Zementwerke, 1956), 23–24. At that time, there was already a "tradition" of industrial espionage as a way of technology transfer. See, e.g., John R. Harris, *Industrial Espionage and Technology Transfer: Britain and France in the Eighteenth Century* (Aldershot, UK: Ashgate Publishing, 1998).

4. For the significance of exhibitions on the development of the concrete industry, see Knut Stegmann, "Introducing Concrete: German Concrete Firms at Exhibitions before 1914," *Engineering History and Heritage* 165, no. 3 (2012): 127–136.

5. A booklet on Coignet's 1867 exhibition was published by Leonard F. Beckwith: *Report on Béton-Coignet, Its Fabrication and Uses: Construction of Sewers, Water-Pipes, Tank, Foundations, Walls, Arches, Buildings, Floors, Terraces, Marine Experiments, etc.* (Washington, D.C.: Government Printing Office, 1868). Frederick A. P. Barnard, president of Columbia College in New York, dealt extensively with Coignet's concrete in his publication on the exhibited machines and industrial processes; see *Paris Universal Exposition, 1867: Report on Machinery and Processes of the Industrial Arts and Apparatus of the Exact Sciences* (New York: D. van Nostrand, 1869), 298–303.

6. See, for instance, the detailed analyses of the German concrete construction company Dyckerhoff & Widmann in Stegmann, *Das Bauunternehmen Dyckerhoff & Widmann*.

7. For an overview of Monier's works, see Jean-Louis Bosc, Jean-Michel Chauveau, Jacques Clément, Jacques Degenne, Bernard Marrey, and Michel Paulin, eds., *Joseph Monier et la Naissance du Ciment Armé* (Paris: Éditions du Linteau, 2001); for Hennebique, with particular focus on his sales strategies, see Gwenaël Delhumeau, *L'invention du béton armé: Hennebique 1890–1914* (Paris: Éditions Norma, 1999), and Gwenaël Delhumeau, Jacques Gubler, Réjean Legault, and Cyrille Simonnet, eds. *Le Béton en Représentation: La Mémoire Photographique de l'entreprise Hennebique 1890–1930* (Paris: Hazan/Institut Français d'Architecture, 1993).

8. Edwin Clarence Eckel, *Portland Cement Materials and Industry in the United States* (Washington, D.C.: Government Printing Office, 1913), 31–35.

9. As late as 1894 the journal *The Engineering Record* presented a German concrete bridge as "one of a class of structures which is clearly destined to come into at least occasional use in America engineering practice as it has already in German practice." "Concrete Bridge over the Danube at Munderkingen," *Engineering Record* 30, no. 23 (1894): 369.

10. In 1912, the professionals Ernest L. Ransome and Alexis Saurbrey wrote in a retrospective that "even as late as 1882, the concrete construction was mainly utilized in foundations and arches suspended between iron beams." *Reinforced Concrete Buildings: A Treatise on the History, Patents, Design and Erection of the Principal Parts Entering into a Modern Reinforced Concrete Building* (New York: McGraw-Hill, 1912), 2.

11. According to Edwin Clarence Eckel, the domestic production of Portland cement in the United States grew from 335,500 barrels a year in 1890 to over 8 million barrels in 1900 and then further to over 76 million barrels in 1910. *Portland Cement Materials and Industry in the United States*, 31.

12. Bill Addis, *Building: 3000 Years of Design Engineering and Construction* (London: Phaidon Press, 2007), 421–427; and Amy E. Slaton, *Reinforced Concrete and the Modernization of American Building, 1900–1930* (Baltimore: Johns Hopkins University Press, 2001), 138–141.

13. These different starting points in the United States and European countries also influenced the development of other industrial branches. See Wolfgang König, "Massenproduktion und Technikkonsum: Entwicklunglinien und Triebkräfte der Technik

zwischen 1880 und 1914," in *Netzwerke, Stahl und Strom. 1840 bis 1914*, ed. Wolfgang König and Wolfhard Weber (Berlin: Ullstein Buchverlage, 2003), 271–274.

14. See, e.g., the descriptions of American machines for concrete mixing incorporated in the first edition of the influential *Handbuch der Ingenieurwissenschaften* and in *Der Portland-Cement und seine Anwendungen im Bauwesen*; Eduard Sonne, "Mörtelmaschinen," in *Handbuch der Ingenieurwissenschaften*, vol. 4.3, *Die Baumaschinen* (Leipzig: Wilhelm Engelmann, 1888), 21, 27; and Friedrich Wilhelm Büsing and Camillo Schumann, eds., *Der Portland-Cement und seine Anwendungen im Bauwesen*, 3rd completely rev. and expanded ed. (Berlin: Kommissions-Verlag Deutsche Bauzeitung, 1905), 282–283. However, there was an interest in American building technique at this time, but that interest was basically limited to constructions in materials other than concrete. American (steel) skyscrapers are, for example, a recurring theme in contemporary German specialist literature.

15. UNESCO World Heritage Committee, Decision: 30 COM 8B.47, "Nominations of Cultural Properties to the World Heritage List (Centennial Hall in Wrocław)," 23 August 2006, http://whc.unesco.org/en/decisions/1010 (accessed 15 July 2012).

16. The city of Wrocław has many different names in different languages. Under German rule, the city was called Breslau. To ensure consistency, this paper invariably uses the current Polish name Wrocław.

17. UNESCO World Heritage Committee, Decision: 30COM 8B.47.

18. For the history of the building and its contexts, see the comprehensive monograph of the Centennial Hall and the Wrocław exhibition ground by Jerzy Ilkosz, *Die Jahrhunderthalle und das Ausstellungsgelände in Breslau: Das Werk Max Bergs*, trans. Beate Störtkuhl (Munich: Oldenbourg Verlag, 2006).

19. The special arrangement of the formwork boards was supposed to positively affect the surface of the unplastered concrete. See Günther Trauer and Willy Gehler, *Die Jahrhunderthalle in Breslau: Berechnung, Konstruktion und Bauausführung* (Berlin: Julius Springer, 1914), 67.

20. Stegmann, *Das Bauunternehmen Dyckerhoff & Widmann*, 167–171.

21. See, e.g., "Breslauer Stadtverordnetenversammlung: Ausstellungs- und Versammlungshalle," *Schlesische Zeitung*, 29 June 1911.

22. The tender asked explicitly for offers both in steel and in concrete construction. See Max Berg and Günther Trauer, "Zweckmäßigkeit von Eisenbeton oder Eisen für monumentale Hochbau-Konstruktionen," *Deutsche Bauzeitung, Mitteilungen über Zement, Beton- und Eisenbetonbau* 10, no. 21 (1913): 163.

23. See Max Berg and Günther Trauer, "Zweckmäßigkeit," 163. The offer for the winning construction in reinforced concrete was 2% more expensive than the cheapest one for steel construction, despite the attempts to optimize the building process. Following criticism from the steel industry, Berg and the Wrocław city construction inspector, Günther Trauer, justified the decision for reinforced concrete because of the better fire protection, lower maintenance costs, and increased longevity of this type of construction.

24. At the turn of the century precast concrete elements were used, for example, for ceiling systems such as the Siegwartbalkendecke in the German Reich. See Friedmar Voormann, "Von der unbewehrten Hohlsteindecke zur Spannbetondecke: Massivdecken zu Beginn des 20. Jahrhunderts," *Beton- und Stahlbetonbau* 100, no. 9 (2005): 842.

25. Willy Gehler, the engineer responsible for the construction of the building at Dyckerhoff & Widmann, explicitly mentions the importance of the transatlantic exchange processes for the building technique used in the Centennial Hall. See Trauer and Gehler, *Die Jahrhunderthalle in Breslau*, 61–62, 66–67.

26. Arjun Appadurai, "Disjuncture and Difference in the Global Cultural Economy," *Theory, Culture and Society* 295, no. 7 (1990): 297.

27. For the change in the perception of space and time, see Stephen Kern, *Culture of Time and Space, 1880–1918*, 2nd ed. (Cambridge, MA: Harvard University Press, 2003).

28. The estate is divided in two archives, the archive of the Deutsches Museum in Munich (NL 050) and that of the Leibniz-Institut für Regionalentwicklung und Strukturplanung in Erkner.

29. See, e.g., Winfried Nerdinger, ed., *Der Architekt Walter Gropius*, 2nd rev. and expanded ed. (Berlin: Gebr. Mann Verlag, 1996), 9–28.

30. Even in Jerzy Ilkosz's extensive monograph, the adaptation of American ideas is only briefly mentioned. *Die Jahrhunderthalle*, 131.

31. For an overview of the current state of research regarding the German Reich, see Stegmann, *Das Bauunternehmen Dyckerhoff & Widmann*, 11–13.

32. Knut Stegmann, "Early Concrete Constructions in Germany—A Review with Special Regard to the Building Company Dyckerhoff & Widmann," in *Proceedings of the Third International Congress on Construction History*, ed. Karl-Eugen Kurrer, Werner Lorenz, and Volker Wetzk (Berlin: Neunplus1 Verlag, 2009), 3:1371. Dyckerhoff & Widmann operated under the name of Lang & Cie. in the early years.

33. There are many sources detailing the global travels of the Dyckerhoffs, including edited diary entries of a world trip in 1898. See Dieter Eißfeldt, *Weltreise 1898: Die Tagebücher von Dr. Wilhelm und Otto Dyckerhoff* (Wiesbaden: Gemeinschaft der Erben Wilhelm G. Dyckerhoff, 2003). The international focus of the Dyckerhoff cement plant corresponded to the growing number of international links in other contemporary fields. See Sebastian Conrad and Jürgen Osterhammel, eds., *Das Kaiserreich transnational: Deutschland in der Welt 1871–1914* (Göttingen: Vandenhoeck & Ruprecht, 2004).

34. Knut Stegmann, "Zu den deutschen Anfängen des Bauens mit Beton: Der Stampfbetonpionier Eugen Dyckerhoff (1844–1924)," *Beton- und Stahlbetonbau* 106, no. 6 (2011): 424.

35. Steel and concrete construction companies competed in more and more tender offers for major building projects in the early twentieth century. One of the early prestigious competitions—the building for the production hall of the Zeppelin airship company in 1908–1909—came to a bad end for the concrete industry. The jury gave all the four awards to steel constructions. See Alfred Haenig, *Luftschiffhallen-Bau: Sammlung moderner Luftschiffhallen-Konstruktionen mit statistischen Berechnungen* (Rostock: Volckmann Nachfolger, 1910), 17–18.

36. Even the well-known, published study trips of the "major" architects only took place later. See, e.g., Erich Mendelsohn, *Amerika: Bilderbuch eines Architekten* (Berlin: Mosse, 1926). The Austrian architect Richard J. Neutra published his well-known book on American architecture in 1927 after he had moved to the United States in 1923. *Wie baut Amerika? Gegenwärtige Bauarbeit. Amerikanischer Kreis* (Stuttgart: Hoffmann, 1927). For other architects, such as Max Berg, study trips to the United States could not be verified. See the resume in Ilkosz, *Die Jahrhunderthalle*, 275–276.

37. Even Walter Gropius discussed, in his well-known article from 1913 on design and construction of industrial buildings in the United States, only formal and constructional issues of American architecture but not details of building technique. "Die Entwicklung moderner Industriebaukunst," in *Die Kunst in Industrie und Handel, Jahrbuch des Deutschen Werkbundes* 2 (1913): 17–22.

38. Franz Widmann, "Bilder zu dem Bericht über eine Studienreise nach America 1909/10," 1910, FA 010/294, Deutsches Museum, Munich. The album was originally part of a detailed report on the study trip. The actual report is no longer traceable.

39. The lecture is printed in two different versions in the proceedings of the association as well as in a supplement of the renowned journal *Deutsche Bauzeitung*. See Deutscher Beton-Verein, *Bericht über die XIV. Haupt-Versammlung des Deutschen Beton-Vereins (E. V.) am 13., 14. und 15. Februar 1911* (Berlin: Tonindustrie-Zeitung, 1911), 288–319; Franz Widmann, "Beton- und Eisenbetonbauten in den Vereinigten Staaten von Nordamerika," pts. 1–2, *Deutsche Bauzeitung, Mitteilungen über Zement, Beton- und Eisenbetonbau* 7, no. 14 (1911): 105–109; no. 15 (1911): 115–117.

40. For a more detailed biography, see Stegmann, *Das Bauunternehmen Dyckerhoff & Widmann*, 132–133.

41. Deutscher Beton-Verein, *Bericht über die XIV. Haupt-Versammlung*, 299.

42. Widmann, "Bilder zu dem Bericht über eine Studienreise nach America 1909/10."

43. Frederick Winslow Taylor and Sanford E. Thompson, eds., *A Treatise on Concrete Plain and Reinforced: Materials, Construction, and Design of Concrete and Reinforced Concrete* (New York: John Wiley, 1905); Frank Gilbreth, *Concrete System* (New York: Engineering News Publishing, 1908).

44. Thus, Franz Widmann ascertained in his study trip that the salaries for typical work in concrete construction were about four times higher than those in Germany. See Deutscher Beton-Verein, *Bericht über die XIV. Haupt-Versammlung*, 289.

45. For an overview of the history of cable cranes on construction sites, see Hans Wettich, "Eine eigenartige Kabelkranform zum Bau der neuen Donaubrücke in Ulm," *Deutsche Bauzeitung* 45, no. 20 (1911): 163–167; "Drahtseilbahnen bei der Ausführung von Ingenieurbauten," *Deutsche Bauzeitung, Mitteilungen über Zement, Beton- und Eisenbetonbau* 8, nos. 17–18 (1911): 133, 135–136, 140–143.

46. Deutscher Beton-Verein, *Bericht über die XIV. Haupt-Versammlung*, 294.

47. Widmann, "Bilder zu dem Bericht über eine Studienreise nach America 1909/10."

48. For example, in railway construction great prefabricated panels were used. See Deutscher Beton-Verein, *Bericht über die XIV. Haupt-Versammlung*, 307–308.

49. Deutscher Beton-Verein, *Bericht über die XIV. Haupt-Versammlung*, 298.

50. No detailed information on Gehler's study has survived. Gehler himself briefly mentioned his trip at the general assembly of the German Concrete Association in 1910 in connection with a molding process he had observed in America. See Deutscher Beton-Verein, *Bericht über die XIII. Haupt-Versammlung des Deutschen Beton-Vereins (E. V.) am 23., 24. und 25. Februar 1910* (Berlin: Tonindustrie-Zeitung, 1910), 81.

51. Trauer and Gehler, *Die Jahrhunderthalle in Breslau*, 61.

52. Trauer and Gehler, *Die Jahrhunderthalle in Breslau*, 66.

53. Trauer and Gehler, *Die Jahrhunderthalle in Breslau*, 74.

54. Pertaining to this saying and its penetration into construction, see Tom Frank Peters, *Time is Money: Die Entwicklung des modernen Bauwesens* (Stuttgart: Julius Hoffmann Verlag, 1981).

55. Trauer and Gehler, *Die Jahrhunderthalle in Breslau*, 61.

56. Hans Hermann Dietrich, "Kabelkrane bei Ausführung von Hochbauten," pt. 2, *Deutsche Bauzeitung, Mitteilungen über Zement, Beton- und Eisenbetonbau* 8, no. 16 (1916): 124.

57. Trauer and Gehler, *Die Jahrhunderthalle in Breslau*, 66.

58. Calculations for the cable cranes and the original application for the construction permit are stored in the Muzeum Architektury we Wrocławiu, Construction Archive, folders 951, 965, and 971.

59. The Bleichert company history is still awaiting a detailed reworking. Thus, at this point the reference is to the 1924 anniversary publication of the company. See Georg Wilhelm Koehler, *Adolf Bleichert & Co. Leipzig: Rückblick und Umschau aus Anlaß des fünfzigjährigen Bestehens am 1. Juli 1924* (Leipzig: Adolf Bleichert, 1924). Some new articles and a comprehensive bibliography of the company's history can be found in Manfred Hötzel and Stefan W. Krieg, eds., *Adolf Bleichert und sein Werk: Unternehmerbiografie, Industriearchitektur, Firmengeschichte*, 2nd rev. ed. (Beucha: Sax-Verlag, 2007).

60. Trauer and Gehler, *Die Jahrhunderthalle in Breslau*, 61–74.

61. Ernest Ransome, for example, developed a prefabrication system for factory construction known as the "Unit System." See Ransome and Saurbrey, *Reinforced Concrete Buildings*, 161–170.

62. Trauer and Gehler, *Die Jahrhunderthalle in Breslau*, 74.

63. An initial overview of the application possibilities offered by pneumatic tools for the concrete industry was given in a lecture by an engineer from a machine factory, A. Beck, at the general assembly of the German Concrete Association in 1911; Deutscher Beton-Verein, *Bericht über die XIV. Haupt-Versammlung*, 320–327. However, Beck mentioned only some attempts in factories for precast concrete elements as references. Apart from these attempts, there were only a few experiments with pneumatic tools on constructions sites, including the use of air rammers in the construction of a bridge in Ulm (Gänstorbrücke, 1910–1912)

and the use of a pneumatic bush hammer in the construction of the concourse hall of the Leipzig Central Station (1909–1911) by Willy Gehler, to rationalize the labor-intensive finish of the visible concrete surfaces in stonemasonry style. For Gänstorbrücke, see Adolf Kleinlogel, "Neue Straßenbrücke über die Donau zwischen Ulm und Neu-Ulm," *Beton und Eisen* 11, no. 2 (1912): 35; for Leipzig Central Station, see Wilhelm Petry, *Betonwerksteine und künstlerische Behandlung des Betons: Entwicklung von den ersten Anfängen der deutschen Kunststeinindustrie bis zur werksteinmäßigen Bearbeitung des Betons* (Munich: Meisenbach Riffarth, 1913), 195.

64. On the occasion of Widmann's lecture at the German Concrete Association, the organization's director, Eugen Dyckerhoff, saw in American building technique "a model for us Germans, to proceed in the application of mechanized operations to reduce costs"; Deutscher Beton-Verein, *Bericht über die XIV. Haupt-Versammlung*, 319.

65. Deutscher Beton-Verein, *Bericht über die XIV. Haupt-Versammlung*, 290–291.

66. Trauer and Gehler, *Die Jahrhunderthalle in Breslau*, 96.

67. "Von der Festhalle in Scheitnig," *Schlesische Zeitung*, 22 December 1912; Trauer and Gehler, *Die Jahrhunderthalle in Breslau*, 74.

68. The lecture held at the general assembly of the German Concrete Association in 1913 appeared in the proceedings, as well as in the professional journal *Armierter Beton* and as an offprint based on the journal article. See Deutscher Beton-Verein, *Bericht über die XVI. Haupt-Versammlung des Deutschen Beton-Vereins (E. V.) am 13., 14. und 15. Februar 1913* (Berlin: Tonindustrie-Zeitung, 1913), 175–196; Günther Trauer and Willy Gehler, "Die Festhalle in Breslau: Berechnung, Konstruktion und Bauausführung," pts. 1–7, *Armierter Beton* 6 (February 1913): 49–66; 6 (April 1913): 150–160; 6 (May 1913): 179–191; 6 (June 1913): 231–240; 7 (January 1914): 8–16, 7 (February 1914): 51–57; 7 (March 1914): 93–101; Trauer and Gehler, *Die Jahrhunderthalle in Breslau*.

69. Trauer and Gehler, *Die Jahrhunderthalle in Breslau*, 61–62, 66–67.

70. Volker Roscher, "Civitas solis—Schöne neue Welt Weimar," in *Zukunft aus Amerika: Fordismus in der Zwischenkriegszeit; Siedlung, Stadt, Raum* (Dessau: Stiftung Bauhaus Dessau, 1995), 68–69.

71. Peter Borscheid, *Naturwissenschaft, Staat und Industrie in Baden (1848–1914)* (Stuttgart: Ernst Klett, 1976), 259–260.

72. Trauer and Gehler, *Die Jahrhunderthalle in Breslau*, 67.

73. Only Hans Schäfer referred briefly in his article on cable cranes in building construction to the social implications of mechanization, albeit only to deny negative effects. He stated rather that "the generous application of machines meant significant health benefits could be expected." "Die Lastenförderung durch Kabelkrane, insbesondere beim Bau der Camsdorfer Brücke," *Beton und Eisen* 8, nos. 18–19 (1914): 361.

74. G.H., "Die Unfallverhütung bei Betonbauten," *Der Grundstein* 25, nos. 48–50 (1912): 596.

75. Well-known projects that were carried out shortly after the Centennial Hall include the Austrian Steyr arms factory, built in 1913–1914, based on plans by Philipp Jakob Manz. See "Mitteilungen aus verschiedenen Fachgebieten: Die Betriebseinrichtungen beim Bau der Waffenfabrik Steyr," *Zeitschrift des Österreichischen Ingenieur- und Architekten-Vereines* 51, no. 47 (1913): 776–777; Kerstin Renz, *Industriearchitektur im frühen 20. Jahrhundert: Das Büro von Philipp Jakob Manz* (Munich: Deutsche Verlags-Anstalt, 2005), 68–71, 171.

76. See H. Ostertag, "Über die Verwendung von Baukabelkranen im Beton- und Eisenbetonbau," pt. 1, *Armierter Beton* 7 (January 1914): 28; and Koehler, *Adolf Bleichert & Co.*, 66, 72. According to the company's published register of completed buildings, Dyckerhoff & Widmann already set up three cable cranes for the excavation work; see Dyckerhoff & Widmann, *Verzeichnis bemerkenswerter Tiefbauausführungen Dywidag. Ausgabe Okt. 1929* (Wiesbaden-Biebrich: Dyckerhoff & Widmann, 1929), group II, 1.

77. Schäfer, "Lastenförderung durch Kabelkrane," 362.

78. Trauer and Gehler, *Die Jahrhunderthalle in Breslau*, 67.

79. However, the Neckar bridge in Stuttgart Bad-Cannstatt completed by Dyckerhoff & Widmann in 1913 showed that the use of compressed-air systems worked out only with optimal utilization. See Heinrich Spangenberg, "Zwei Betonbauten vom Stuttgarter Bahnhof-Umbau: Der Doppeltunnel durch den Rosenstein und die viergleisige Eisenbahnbrücke über den Neckar," pt. 2, *Deutsche Bauzeitung, Mitteilungen über Zement, Beton- und Eisenbetonbau* 11, no. 11 (1914): 84.

80. Countless publications give the site at Wrocław as a template for the use of cable cranes in building construction without any reference to the American models that it was based on. See, e.g., Max Buhle, "Neuzeitliche Kabelkrane und ihre Anwendung auf das Bauwesen," pt. 3, *Deutsche Bauzeitung* 47, no. 82 (1913): 746, 748; Dietrich, "Kabelkrane bei Ausführung von Hochbauten," 121–123; Ostertag, "Verwendung von Baukabelkranen," 27–32, 69–70; see also Max Buhle, "Krane für Massentransport," in *Lexikon der gesamten Technik und ihrer Hilfswissenschaften*, ed. Otto Lueger, 2nd completely rev. ed. (Stuttgart: Deutsche Verlags-Anstalt, 1914), 456. The Bleichert company itself also presented the site without any reference to its American forerunners in its anniversary publication from 1924; see Koehler, *Adolf Bleichert & Co.*, 74, 81.

Bibliography

Addis, Bill. *Building: 3000 Years of Design Engineering and Construction*. London: Phaidon Press, 2007.

Appadurai, Arjun. "Disjuncture and Difference in the Global Cultural Economy." *Theory, Culture and Society* 295, no. 7 (1990): 295–310.

Barnard, Frederick A. P. *Paris Universal Exposition, 1867: Report on Machinery and Processes of the Industrial Arts and Apparatus of the Exact Sciences*. New York: D. van Nostrand, 1869.

Beckwith, Leonard F. *Report on Béton-Coignet, Its Fabrication and Uses: Construction of Sewers, Water-Pipes, Tank, Foundations, Walls, Arches, Buildings, Floors, Terraces, Marine Experiments, etc.* Washington, D.C.: Government Printing Office, 1868 (Paris Universal Exhibition, 1867. Reports of the United States Commissioners).

Berg, Max, Papers. NL 050. Deutsches Museum, Munich.

————, Papers. Leibniz-Institut für Regionalentwicklung und Strukturplanung, Erkner, Germany.

Berg, Max, and Günther Trauer. "Zweckmäßigkeit von Eisenbeton oder Eisen für monumentale Hochbau-Konstruktionen." *Deutsche Bauzeitung, Mitteilungen über Zement, Beton- und Eisenbetonbau* 10, no. 21 (1913): 163, 165.

Bleichert. Company Archive. No. 20781. Staatsarchiv Leipzig.

Bonner Portland-Zementwerke. *Ein Jahrhundert Bonner Zement: 1856–1956*. Bonn: Bonner Portland-Zementwerke, 1956.

Borscheid, Peter. *Naturwissenschaft, Staat und Industrie in Baden (1848–1914)*. Stuttgart: Ernst Klett, 1976.

Bosc, Jean-Louis, Jean-Michel Chauveau, Jacques Clément, Jacques Degenne, Bernard Marrey, and Michel Paulin, eds. *Joseph Monier et la Naissance du Ciment Armé*. Paris: Éditions du Linteau, 2001.

"Breslauer Stadtverordnetenversammlung: Ausstellungs- und Versammlungshalle." *Schlesische Zeitung,* 29 June 1911.

Buhle, Max. "Krane für Massentransport." In *Lexikon der gesamten Technik und ihrer Hilfswissenschaften,* ed. Otto Lueger, pp. 449–457. 2nd completely rev. ed. Stuttgart: Deutsche Verlags-Anstalt, 1914.

————. "Neuzeitliche Kabelkrane und ihre Anwendung auf das Bauwesen." Pts. 1–3. *Deutsche Bauzeitung* 47, no. 79 (1913): 716–720; 47, no. 81 (1913): 734–736; 47, no. 82 (1913): 746–748.

Büsing, Friedrich Wilhelm, and Camillo Schumann, eds. *Der Portland-Cement und seine Anwendungen im Bauwesen*. 3rd completely rev. and expanded ed. Berlin: Kommissions-Verlag Deutsche Bauzeitung, 1905.

"Concrete Bridge over the Danube at Munderkingen." *Engineering Record* 30, no. 23 (1894): 369.

Conrad, Sebastian, and Jürgen Osterhammel, eds. *Das Kaiserreich transnational: Deutschland in der Welt 1871–1914*. Göttingen: Vandenhoeck & Ruprecht, 2004.

Delhumeau, Gwenaël. *L'invention du béton armé: Hennebique 1890–1914*. Paris: Éditions Norma, 1999.

Delhumeau, Gwenaël, Jacques Gubler, Réjean Legault, and Cyrille Simonnet, eds. *Le Béton en Représentation: La Mémoire Photographique de l'entreprise Hennebique 1890–1930*. Paris: Hazan/Institut Français d'Architecture, 1993.

Deutscher Beton-Verein. *Bericht über die XIII. Haupt-Versammlung des Deutschen Beton-Vereins (E. V.) am 23., 24. und 25. Februar 1910*. Berlin: Tonindustrie-Zeitung, 1910.

————. *Bericht über die XIV. Haupt-Versammlung des Deutschen Beton-Vereins (E. V.) am 13., 14. und 15. Februar 1911*. Berlin: Tonindustrie-Zeitung, 1911.

————. *Bericht über die XVI. Haupt-Versammlung des Deutschen Beton-Vereins (E. V.) am 13., 14. und 15. Februar 1913*. Berlin: Tonindustrie-Zeitung, 1913.

Dietrich, Hans Hermann. "Kabelkrane bei Ausführung von Hochbauten." Pts. 1–2. *Deutsche Bauzeitung, Mitteilungen über Zement, Beton- und Eisenbetonbau* 8, no. 14 (1916): 105–107; no. 16 (1916): 121–124.

Dyckerhoff & Widmann. *Verzeichnis bemerkenswerter Tiefbauausführungen Dywidag. Ausgabe Okt. 1929*. Wiesbaden-Biebrich: Dyckerhoff & Widmann, 1929.

Dyckerhoff & Widmann AG (DYWIDAG). Company Archive. FA 010. Deutsches Museum, Munich.

Eckel, Edwin Clarence. *Portland Cement Materials and Industry in the United States*. Washington, D.C.: Government Printing Office, 1913.

Eißfeldt, Dieter. *Weltreise 1898: Die Tagebücher von Dr. Wilhelm und Otto Dyckerhoff*. Wiesbaden: Gemeinschaft der Erben Wilhelm G. Dyckerhoff, 2003.

Gilbreth, Frank. *Concrete System*. New York: Engineering News Publishing, 1908.

Gropius, Walter. "Die Entwicklung moderner Industriebaukunst." In *Die Kunst in Industrie und Handel. Jahrbuch des Deutschen Werkbundes* 2 (1913): 17–22.

H., G. "Die Unfallverhütung bei Betonbauten." Pts. 1–3. *Der Grundstein* 25, no. 48 (1912): 578–580 no. 49 (1912): 596–597; no. 50 (1912): 608–609.

Haegermann, Gustav. "Vom Caementum zum Zement." In *Vom Caementum zum Zement,* ed. Günter Huberti, pp. 3–72. Wiesbaden: Bauverlag, 1964.

Haenig, Alfred. *Luftschiffhallen-Bau: Sammlung moderner Luftschiffhallen-Konstruktionen mit statistischen Berechnungen*. Rostock: Volckmann Nachfolger, 1910.

Harris, John R. *Industrial Espionage and Technology Transfer: Britain and France in the Eighteenth Century*. Aldershot, UK: Ashgate Publishing, 1998.

Hötzel, Manfred, and Stefan W. Krieg, eds. *Adolf Bleichert und sein Werk: Unternehmerbiografie, Industriearchitektur, Firmengeschichte*. 2nd rev. ed. Beucha: Sax-Verlag, 2007.

Ilkosz, Jerzy. *Die Jahrhunderthalle und das Ausstellungsgelände in Breslau: Das Werk Max Bergs*. Translated by Beate Störtkuhl. Munich: Oldenbourg Verlag, 2006.

Kern, Stephen. *Culture of Time and Space, 1880–1918*. 2nd ed. Cambridge, MA: Harvard University Press, 2003.

Kleinlogel, Adolf. "Neue Straßenbrücke über die Donau zwischen Ulm und Neu-Ulm." *Beton und Eisen* 11, no. 2 (1912): 35–36.

Koehler, Georg Wilhelm. *Adolf Bleichert & Co. Leipzig: Rückblick und Umschau aus Anlaß des fünfzigjährigen Bestehens am 1. Juli 1924*. Leipzig: Adolf Bleichert, 1924.

König, Wolfgang. "Massenproduktion und Technikkonsum: Entwicklunglinien und Triebkräfte der Technik zwischen 1880 und 1914." In *Netzwerke, Stahl und Strom. 1840 bis 1914,* ed. Wolfgang König and Wolfhard Weber, pp. 265–552. Berlin: Ullstein Buchverlage, 2003.

———. *Technikgeschichte: Eine Einführung in ihre Konzepte und Forschungsergebnisse*. Stuttgart: Franz Steiner Verlag, 2009.

Mendelsohn, Erich. *Amerika: Bilderbuch eines Architekten*. Berlin: Mosse, 1926.

"Mitteilungen aus verschiedenen Fachgebieten: Die Betriebseinrichtungen beim Bau der Waffenfabrik Steyr." *Zeitschrift des Österreichischen Ingenieur- und Architekten-Vereines* 51, no. 47 (1913): 776–777.

Muzeum Architektury we Wrocławiu. Construction Archive, folders 951, 965, 971.

Nerdinger, Winfried, ed. *Der Architekt Walter Gropius*. 2nd rev. and expanded ed. Berlin: Gebr. Mann Verlag, 1996.

Neutra, Richard J. *Wie baut Amerika? Gegenwärtige Bauarbeit. Amerikanischer Kreis*. Stuttgart: Hoffmann, 1927.

Ostertag, H. "Über die Verwendung von Baukabelkranen im Beton- und Eisenbetonbau." Pts. 1–2. *Armierter Beton* 7 (January 1914): 27–32; 7 (February 1914): 67–70.

Peters, Tom Frank. *Time is Money: Die Entwicklung des modernen Bauwesens*. Stuttgart: Julius Hoffmann Verlag, 1981.

Petry, Wilhelm. *Betonwerksteine und künstlerische Behandlung des Betons: Entwicklung von den ersten Anfängen der deutschen Kunststeinindustrie bis zur werksteinmäßigen Bearbeitung des Betons*. Munich: Meisenbach Riffarth, 1913.

Ransome, Ernest L., and Alexis Saurbrey. *Reinforced Concrete Buildings: A Treatise on the History, Patents, Design and Erection of the Principal Parts Entering into a Modern Reinforced Concrete Building*. New York: McGraw-Hill, 1912.

Renz, Kerstin. *Industriearchitektur im frühen 20. Jahrhundert: Das Büro von Philipp Jakob Manz*. Munich: Deutsche Verlags-Anstalt, 2005.

Roscher, Volker. "Civitas solis—Schöne neue Welt Weimar." In *Zukunft aus Amerika: Fordismus in der Zwischenkriegzeit; Siedlung, Stadt, Raum*, pp. 64–91. Dessau: Stiftung Bauhaus Dessau, 1995.

Schäfer, Hans. "Die Lastenförderung durch Kabelkrane, insbesondere beim Bau der Camsdorfer Brücke." *Beton und Eisen* 8, nos. 18–19 (1914): 361–364.

Slaton, Amy E. *Reinforced Concrete and the Modernization of American Building, 1900–1930*. Baltimore: Johns Hopkins University Press, 2001.

Sonne, Eduard. "Mörtelmaschinen." In *Handbuch der Ingenieurwissenschaften*, vol. 4.3, *Die Baumaschinen*, pp. 133–248. Leipzig: Wilhelm Engelmann, 1888.

Spangenberg, Heinrich. "Zwei Betonbauten vom Stuttgarter Bahnhof-Umbau: Der Doppeltunnel durch den Rosenstein und die viergleisige Eisenbahnbrücke über den Neckar." Pts. 1–4. *Deutsche Bauzeitung, Mitteilungen über Zement, Beton- und Eisenbetonbau* 11, no. 10 (1914): 73–78; no. 11 (1914): 82–84; no. 12 (1914): 89–93; no. 13 (1914): 97–102.

Stegmann, Knut. *Das Bauunternehmen Dyckerhoff & Widmann: Zu den Anfängen des Betonbaus in Deutschland 1865–1918*. Tübingen and Berlin: Wasmuth Verlag, 2014.

———. "Early Concrete Constructions in Germany—A Review with Special Regard to the Building Company Dyckerhoff & Widmann." In *Proceedings of the Third International Congress on Construction History*, ed. Karl-Eugen Kurrer, Werner Lorenz, and Volker Wetzk, vol. 3, pp. 1371–1378. Berlin: Neunplus1 Verlag, 2009.

———. "Introducing Concrete: German Concrete Firms at Exhibitions before 1914." *Engineering History and Heritage* 165, no. 3 (2012): 127–136.

———. "Zu den deutschen Anfängen des Bauens mit Beton: Der Stampfbetonpionier Eugen Dyckerhoff (1844–1924)." *Beton- und Stahlbetonbau* 106, no. 6 (2011): 415–424.

Taylor, Frederick Winslow, and Sanford E. Thompson, eds. *A Treatise on Concrete Plain and Reinforced: Materials, Construction, and Design of Concrete and Reinforced Concrete*. New York: John Wiley, 1905.

Trauer, Günther, and Willy Gehler. "Die Festhalle in Breslau: Berechnung, Konstruktion und Bauausführung." Pts. 1–7. *Armierter Beton* 6 (February 1913): 49–66; 6 (April 1913): 150–160; 6 (May 1913): 179–191; 6 (June 1913): 231–240; 7 (January 1914): 8–16; 7 (February 1914): 51–57; 7 (March 1914): 93–101.

———. *Die Jahrhunderthalle in Breslau: Berechnung, Konstruktion und Bauausführung*. Berlin: Julius Springer, 1914.

UNESCO World Heritage Committee. Decision: 30 COM 8B.47. "Nominations of Cultural Properties to the World Heritage List (Centennial Hall in Wroclaw)." 23 August 2006. http://whc.unesco.org/en/decisions/1010 (accessed 15 July 2012).

"Von der Festhalle in Scheitnig." *Schlesische Zeitung*, 22 December 1912.

Voormann, Friedmar. "Von der unbewehrten Hohlsteindecke zur Spannbetondecke: Massivdecken zu Beginn des 20. Jahrhunderts." *Beton- und Stahlbetonbau* 100, no. 9 (2005): 836–846.

Wettich, Hans. "Drahtseilbahnen bei der Ausführung von Ingenieurbauten." Pts. 1–2. *Deutsche Bauzeitung, Mitteilungen über Zement, Beton- und Eisenbetonbau* 8, no. 17 (1911): 133, 135–136; no. 18 (1911): 140–143.

———. "Eine eigenartige Kabelkranform zum Bau der neuen Donaubrücke in Ulm." *Deutsche Bauzeitung* 45, no. 20 (1911): 163–167.

Widmann, Franz. "Beton- und Eisenbetonbauten in den Vereinigten Staaten von Nordamerika." Pts. 1–2. *Deutsche Bauzeitung, Mitteilungen über Zement, Beton- und Eisenbetonbau* 8, no. 14 (1911): 105–109; no. 15 (1911): 115–117.

———. "Bilder zu dem Bericht über eine Studienreise nach America 1909/10." 1910. FA 010/294. Deutsches Museum, Munich.

CHAPTER 2

Heating the Groves
The Globalization of Agricultural Technology and the Acclimatization of Citrus in the USSR, 1928–1936

Johanna Conterio
Postdoctoral Research Fellow

Birkbeck College
University of London
London, UK

Recent research in the history of technology of the Eastern Bloc has highlighted the relative opening of technology transfer between the East and West during the Cold War after Stalin. As Karen Johnson Freeze has found, transfer of technology and culture did not flow only in one direction, from the West to the East. Technology transfer was a two-way process. Ideas and technology produced in socialist countries were adopted and assimilated on both sides of the Iron Curtain.[1] György Péteri has even questioned the viability of the concept of the "iron curtain" in work about the exchange of technology and culture.[2]

However, the understanding of the mechanism of technology transfer during the period of most rapid technological change in the Soviet Union, the years of the First and Second Five-Year Plans (1928–1933, 1933–1937), when the party-state led a course of crash industrialization and the collectivization of agriculture, has been left largely unchallenged since the end of the Cold War. The historiography continues to maintain that the transfer of technology between systems after 1931, following a period of iconic industrial development based on foreign contracts and executed by foreign engineers, was deterred by ideological barriers to foreign technology and to interactions with the West. After 1931, Soviet investment in foreign technology was dramatically reduced.[3] As the Stalinist system consolidated power and promoted an increasingly xenophobic policy focused on the development of domestic industry, science, and technology, imports suggested the failure of the autarkic project. Yet, as this chapter will suggest, the globalization of technology under Stalin was thwarted by economic concerns as well, particularly in the low-priority realm of agricultural technology.[4] This chapter takes the case of an agricultural technology to support new citriculture in the Soviet south, orchard heaters, to illustrate that decisions about importing technology into the Soviet Union were based on economic as well as ideological criteria, although the decision was publicly framed in terms of ideological orthodoxy. Here the focus will be not on whether the technology in question—orchard heaters—was imported, but rather on the decision-making process leading up to a decision, weighing the relative influence of ideological and economic concerns.

The "American" Method

In 1929, California led the world in orange production and in the development of citrus technology. Although heaters were occasionally used to heat deciduous orchards, in California orchard heating was a practice largely concentrated in citriculture. The citrus orchards of California were threatened by frequent frosts, and because citrus trees bore fruit during the winter and early spring, frosts could destroy an entire citrus harvest of nearly mature fruits and might also damage trees, resulting in reduced crops in the following year. Deciduous trees were less prone to damage because they blossomed and grew later in the spring. The use of orchard heaters increased as citriculture took on an industrial character in the 1910s and 1920s and the potential economic impact of failed crops due to frost grew. The use of orchard heaters grew particularly quickly during a series of cold winters in the early 1920s. In 1925, it was estimated that 1,500,000 orchard heaters were in use statewide, extending frost protection over about 30,000 acres.[5]

In the 1920s and 1930s, citrus was being acclimatized to countries all over the world. The orange-growing countries of the world included, in 1938, Jamaica, Cuba, Puerto Rico and the British West Indies, Mexico, Honduras, Costa Rica and Ecuador, Argentina and Brazil, Mozambique, South Africa, Southern Rhodesia, Egypt, France, Spain, Turkey, India, China, Japan, the Philippines, Australia, and New Zealand. As the director of the Agricultural Experiment Station at the University of California, Berkeley, E. D. Merrill wrote that the California method of protecting groves against damaging temperatures through the use of orchard heating was of international interest. Specialists at the Agricultural Experiment Station corresponded with colleagues in India, Egypt, Mandate Palestine, Morocco, and Algeria.[6] Furthermore, the work of the Agricultural Experiment Station had an even broader geographical reach through its publications, the *Bulletin of the Agricultural Experiment Station*. Information from California about citricultural technology spread to a global reach.

In 1925, a group of specialists published a special issue of the *Bulletin* dedicated to the topic of orchard heating, *Orchard Heating in California*, citing international interest as their motivation. The issue provided an overview of a survey of orchard-heating practices, costs, and results among

California farmers, conducted by Robert Hodgson, associate professor of subtropical horticulture and associate citriculturist; Warren R. Schoonover, an extension specialist in citriculture; and Floyd D. Young, a meteorologist at the U.S. Weather Bureau in charge of the Fruit Frost Service.

As Schoonover, Hodgson, and Young described, an important technological breakthrough in orchard-heating technology came with the introduction of the first orchard heater fueled by oil. Early experiments with the use of oil to heat orchards were conducted by a California entrepreneur, Charles Froude, in the 1890s.[7] At that time, most orchard heaters in use in California were fueled by coal. These were difficult to light and extinguish, leading to high labor costs and wasted fuel. Fuel-burning orchard heaters also supplemented the practice of burning wet straw and compost in the groves to raise temperatures, called "smudging."[8]

The industrial production of oil-fueled orchard heaters came about two decades later, the "lard pail heater." Also known as the "Fresno pot" or the "California oil pot," it was made of a simple pail with sloping sides, a flame spreader (wick), a "spider" (a device to slow burning), and a cover. In 1925, the heaters were produced by a number of manufacturers, including Bolton and Canco, but sizes were standardized: holding five or ten quarts of oil. The heaters burned for about 3 to 3.5 hours without the spider and 9 to 10 hours with the spider.[9]

The shortcomings of the lard pail heater were numerous. The heaters did not burn all night, and if workers failed to refill them, fruits were left exposed to the coldest hours of the night before sunrise.[10] Lard pail heaters left a smoky smudge on citrus fruits. These smudges could be cleaned, but the process was labor intensive, and sometimes smudges nevertheless lowered the grade of the fruits. Because of their relatively low capacity, a large number of heaters were required. Also, although they were effective at raising the air temperature a few degrees, they were not powerful enough for effective use in colder regions.[11]

To address these shortcomings, new technologies were developed (Figure 1). As *Orchard Heating in California* noted, the "low stack distilling type oil heaters," "medium stack down draft distilling oil heaters," and "tall stack distilling oil heaters" all burned through the night without requiring refilling. The "large capacity heaters with burner and oil reservoir separate" were far more powerful than earlier models and had the advantage of drawing oil automatically from an attached reserve. But the California specialists had economic rationality in mind, recommending the careful calculation of fuel and heating needs based on local climate and of the costs that would be associated with their provision. Only if the benefits outweighed the costs should farms invest in technology, and the most advanced technology was not necessarily the best choice. Indeed, the specialists recommended the continued use of simpler, older technologies such as the "briquette burners" and "coal burners," particularly as supplemental heaters to be used during particularly cold spells.

Soviet Citrus

The Stalinist plan to modernize agriculture rested on the regional specialization of agriculture and large-scale farming to maximize agricultural production, minimize labor inputs, and replace spontaneous and varied local crops with rationally selected, highly profitable monocultures.[12] These

1.
Oil-burning orchard heaters were used extensively by California citrus growers beginning in the early twentieth century in efforts to keep their trees and fruit from freezing. The heaters produced not only warmth but thick black smoke. This photo was taken in 1949 about the time that such heaters were outlawed and other frost-prevention methods took over. Photo from The David Boule Collection.

policies were implemented through the forceful collectivization of agriculture, beginning in 1928. From 1929, the development of subtropical agriculture in the Soviet Union was encouraged. That year, the All-Union Committee on the Subtropics was established under the Soviet of Work and Defense and launched a publication, *Sovetskie subtropiki* (The Soviet Subtropics). *Sovetskie subtropiki* encouraged subtropical agriculture and attempted to mobilize research institutions in the region. In the early 1930s, some state farms were also set up for subtropical agriculture.[13]

In 1933, a more ambitious plan for the development of the Soviet subtropics was announced on the pages of the newspaper *Pravda*. The plan introduced a territorial concept for the new agriculture, the "subtropics." The subtropics stretched along the Black Sea coast from the Russian city of Sochi in the north to the Turkish border in the south. Among the subtropical crops prioritized was citrus. Stalin gave the assignment that by 1937 the harvest of citrus fruits from the Soviet subtropics would be no less than 500 million fruits, up from a reported 100 million fruits recorded in the 1933 harvest. The experts of the newly established Administration of Subtropical Agriculture of the People's Commissariat (Ministry) of Agriculture were aware that the climate of the Black Sea coastal region and the idiosyncrasies of its landscape, which was marked by swamps and ravines and prone to earthquakes and landslides, would pose a challenge to the new agriculture. But they were confident that Soviet agricultural science would prove up to the task.

As the northernmost area of the newly declared citrus-growing region, Sochi faced the greatest climatic obstacles to citrus growing. The region was prone to frequent frosts. Agriculture was overwhelmingly focused on corn and tobacco in the region at the time.[14] But local scientists at the Sochi Experimental Station for Southern and Subtropical Agriculture, an institute of the Commissariat of Agriculture, diligently began experiments with citrus in the late 1920s, turning from a focus on plums and the introduction of vegetables (such as cauliflower from Germany). In 1930, the station had 1,000 citrus trees in an experimental orchard.[15]

In the cold winter of 1928–1929, the Sochi Experimental Station experimented with the use of orchard heaters.[16] Very little information is available about how and where these heaters were produced. A photograph of the Sochi orchard heaters published in 1935 indicates that the station used orchard heaters of the lard pail type, which were referred to as "oil heater-buckets" (*neftianykh grelok-veder*). But what the archival evidence makes clear is that they were not models from the United States.[17]

The Sochi Experimental Station promoted the development of tangerine cultivation in the region under its jurisdiction.[18] This territory stretched from the border with Abkhazia at the Psou River in the south to the settlement of Kabardinka, about 250 km north along the Black Sea coast from Psou. The territory made up 645,473 hectares (1,594,318 acres), of which 75,409 hectares (337,078 acres) was farmland. In 1933, it had a population of 123,730, of which 53,142 lived in agricultural villages and collective and state farms.[19]

Evidence suggests that Sochi was the first location in the Soviet Union where specialists employed oil-burning orchard heaters and possibly any form of open-air orchard heating.[20] In 1935, N. A. Baril'chenko, a specialist at the Sochi Experimental Station, wrote in the journal of the station that orchard heaters were used first in Sochi: "A significant attack of tangerines was observed in the cold winter 1928–1929, when all subtropical plantings strongly suffered from frost reaching –11° Celsius. The heating of tangerines was not adopted in a single plot except at the experimental station."[21] The regional origins of the practice was confirmed by N. K. Ianushevskii, director of the All-Union Scientific Research Institute of the Humid Subtropics, who emphasized in a report in *Sovetskie subtropiki* in 1934 that interest in the technology had been initiated at the local level, in the regions: "The question of the use of American methods of protection from frost in our conditions began to be worked out by the scientific-research institutions of the subtropical regions from 1930."[22] Sochi was credited with introducing the scientific study of the practice to the Soviet subtropics. In another article published in 1934 in *Sovetskie subtropiki*, G. T. Selianinov, a Leningrad professor, noted that the earliest experiments with the technology were conducted in Sochi.[23]

Both the director of the Sochi Experimental Station, Iakhontov, and his deputy for fruit culture, Boris Aznin, recognized the economic potential of the orchard-heating technology for Soviet citriculture.[24] They had good reason to. The 1928–1929 experiments were successful. The station reported that the oil heaters raised the temperatures in the groves by 2°C–4.5°C, and they conducted a controlled study of the practice, comparing areas where heaters were used with unheated groves. In the areas where no heating was used, 40% of the trees suffered loss of leaves and frost damage, and the grove yielded no harvest. In the heated portion, on the other hand, there was even a small harvest.

Not only had their experiments been successful, but early and widespread interest in the new orchard-heating technology poured into the station. These queries arrived from throughout the subtropical region, beyond the region in the nominal jurisdiction of the station, quite like what had occurred at the Agricultural Experiment Station in California in the early 1920s. In November 1929, for example, the director of the State Farm Psyrtskha in Abkhazia, to the south of the region covered by the station, wrote to the Sochi Experimental Station asking the scientists to share the results of their frost prevention measures of 1928–1929. Word traveled fast, possibly through social or scientific networks, possibly through administrators, possibly through previous correspondence with the Sochi Experimental Station.[25]

The specialists at the Sochi Experimental Station longed to study the most advanced orchard-heating technology from the United States. But as of 1930, Boris Aznin reported the station had not been able to buy any American technology. Aznin wrote that the reasons for this were economic: the Sochi Experimental Station did not have the funds to purchase the heaters.[26] But the Sochi Experimental Station actively set about the task of getting the most advanced models of orchard heaters from California for study. Lacking the resources to buy the heaters, Aznin promoted the development and use of the technology among the better-funded state farms. His strategy was to encourage state farmers to demand the technology from their administrative superiors (Figure 2). As his letters made clear, his own aim was to acquire a few prototypes for study at the Sochi Experimental Station, to be used to develop Soviet models. He acquired information about these advanced technologies from the Agricultural Experiment Station at the University of California, Berkeley. Indeed, the Sochi Experimental Station received issues of the *Bulletin* from their Californian colleagues.

In a 1930 letter to the director of the State Farm imeni Lenina, Aznin described the successes of the station with the technology and encouraged the state farm to agitate their superiors to import the technology. He wrote,

> We hope that you will be interested in the speedy development of our work to improve the methods of protection and will help us to acquire models of heaters used in California. To this end, we are including here the names, forms, specifications and prices for the main types of heaters that we hope to get a copy of.[27]

Iakhontov also attempted to gather models for study from domestic producers. Iakhontov wrote in January 1930 to a specialist in heating technology, P. E. Shteinlekhner, in the city of Riazan, requesting recommendations for domestic heaters suited to orchard use. He asked that two suitable "ovens" be sent to the station for experimental work. He requested a heater that would raise the temperature of the air by 5°–10° and could be placed between trees. He hinted that the heaters might become a cause attracting investment in the future: "We add that the issue of heating systems for the tangerines and sheds is currently a high priority. If their use is successful

Отепление американскими нефтяными грелками взрослого насаждения лимонов в совхозе „Псырцха". (Фото Суровцева).

2.
American orchard heaters at the State Farm Psyrtskha. From *Sovetskie subtropiki*, 1934. Courtesy of the Botany Libraries, Harvard University.

we can expect their mass application."[28] The state had a record of importing foreign technology for broad domestic use in the Soviet Union in strategic industries during the First Five-Year Plan. But neither Aznin nor Iakhontov anticipated that orchard heaters would be included in this elite list, hoping only for the importation of prototypes for study.

In the Stalinist 1930s, it was extremely difficult, if not impossible, for a local enterprise to legally import technology directly from abroad. The state controlled import and export markets through its agencies abroad.[29] From 1931, when funding of imports collapsed, there was a rapid decline in official Soviet enthusiasm for American technology.[30]

Michurinist Opposition

A. M. Lezhava, the director of the Administration of Subtropical Agriculture of the People's Commissariat of Agriculture, was aware that the climate of the Black Sea coastal region and the idiosyncrasies of its landscape, which was marked by swamps and ravines and prone to earthquakes and landslides, would pose a challenge to the new subtropical agriculture. The region was climatically diverse. There had been debates throughout the 1920s about whether Sochi was even subtropical. But Lezhava and the other experts of the administration were confident that

Soviet agricultural science and engineering would prove up to the task.[31] The Administration of Subtropical Agriculture treated the development of the subtropics as a challenge for Soviet agricultural science.

The Administration of Subtropical Agriculture led an aggressive expansion of research on citrus cultivation in the research institutions of the Commissariat of Agriculture. To increase co-ordination between the center and research institutions, a new, unified institute for the study of the subtropics, the All-Union Scientific Research Institute of the Humid Subtropics, was opened in April 1933. The mandate of the institute was to unify and rationalize the plan for all research on the subtropics in the research institutions of the USSR, eliminating parallelism and coordinating the research agenda.[32] It also increased its control over citriculture research in the institutions in its jurisdiction. The institute was opened at a peak in state confidence in agricultural research as a route toward rapid agricultural progress; in 1932–1933 there were almost 1,300 institutions, employing near 26,000 specialists, dedicated to agricultural research.[33]

The position of the Administration of Subtropical Agriculture and the All-Union Scientific Research Institute of the Humid Subtropics increasingly focused on overcoming climatic barriers to the introduction of citrus through Michurinist biology (methods focused on breeding new species), which would be engaged to create hybrid species of citrus that were frost resistant.[34] Indeed, the creation of the Institute of Humid Subtropics coincided with the rise of Michurin in late 1933, when his institute was broadly expanded and given increased jurisdiction over agricultural research.[35]

The All-Union Scientific Research Institute of the Humid Subtropics led by example. The institute reported on breeding experiments that it had undertaken in 1933, which were particularly focused on the creation of frost-resistant breeds of citrus. As the institute reported, it had conducted 10,000 cross-breeding experiments with tangerines, 2,000 with lemons, and 1,000 with other citrus, all undertaken with the aim of creating frost-resistant, high-yielding, and high-quality breeds.[36] Moreover, the institute began in 1933 to study vernalization, the technique introduced by Lysenko for treating seeds before planting.[37]

The All-Union Scientific Research Institute of the Humid Subtropics became increasingly interventionist in the research agendas of subordinate institutions. The institute unveiled an ambitious plan for coordinated research in all the institutes involved in subtropical agricultural research in 1934 that was approved by the Committee of Subtropical Agriculture.[38] The plan had a decided emphasis on Michurinist "agrobiology." Topics such as vernalization were assigned. But the plan still allocated resources for other types of research, such as the selection of already existing frost-resistant species for introduction, acclimatization, and propagation in the Soviet subtropics; experiments with orchard technology; and research into pests and diseases. The commitment to Michurinism and Lysenkoism would become more pronounced from 1935. But the center consistently focused on laboratory-based scientific research more than on the development of new agricultural technology (tractors, planters, hoes, insulation, etc.). But was the reason for this focus primarily ideological or economic?

In 1935, the aggressive push for Michurinist breeding techniques and the rise of Lysenko threatened the use of orchard heating in Sochi and, indeed, in the entire Soviet subtropics. Until 1935, a turning point in Soviet biology, Michurinist research coexisted with a broad research program. However, this research was increasingly in conflict with directives from the center. In 1935, the Administration of Subtropical Agriculture made an effort to control local research initiatives. That year, it instructed the Sochi Experimental Station to set up a scientific-technical council, which the administration would approve.

With this turn in agricultural policy, the center sought out new heroes. In Sochi, the new policy lifted up the fate of a fruit breeder at the Sochi Experimental Station, F. M. Zorin. On 13 October 1935, an article in *Pravda* dedicated to the Sochi Experimental Station described work underway at the station to create citrus hybrids. The aim of this work was to develop a frost-resistant tangerine, suited to mass distribution in the whole Sochi region and even farther north. The article pointed out that in Sochi tangerine trees suffered from frost-related damage every four years and every twelve years were destroyed by frost. Writing about Zorin, the article reiterated the central vision for agricultural science in the subtropics: "To find, build that type of tangerine, which would not fear any caprices of the Sochi climate and at the same time would give large harvests of high-quality fruit—that is what Zorin and his group are persistently working on."[39] Zorin was lifted from relative obscurity at the station among scientists and practical researchers to a national hero. In 1935, the scientific research plan at the local station, approved by the Administration of Subtropical Agriculture, saw an exponential expansion of research dedicated to breeding. New agricultural equipment remained underfunded and underprioritized in research plans.

In 1936, the Sochi Experimental Station scientific council held an assembly dedicated to the anniversary of the death of Michurin.[40] The gathering approved the goal to promote Michurinist methods to create a new breed of frost-resistant subtropical culture. As the assembly resolution stated, "Approve the proposal of the Department of Selection and Species Research to create new sorts of frost-resistant subtropical agriculture using the methods of Michurin by the year 1940."[41]

Local officials, impressed by the indications coming from the center to focus local research initiatives exclusively on Michurinist biology and eager to demonstrate their loyalty to the center, denounced the use of orchard heaters, calling their use tantamount to proclaiming the region unsuitable for citrus. The Plenipotentiary of the Central Executive Committee for the Sochi-Matsesta Health Resort, Denis Metelev, openly protested the use of orchard heaters, which he considered an expression of skepticism about the viability of citrus crops in the Sochi region and even sabotage. In a talk given in March 1935, Metelev argued,

> In Sochi even now the perspective reigns that the natural climatic conditions of our region are not suited to tangerine agriculture. This point of view was supported by the experimental station, as well as the former director of the Regional Executive Committee and a whole row of other comrades, who said that frost could destroy our groves, that the land, climate, humidity would not allow the development of this culture. I think that we will talk

a lot more about this question. Our experimental station, which came to us in 1934 with empty hands, has its tangerine grove, but their attempts in this regard were limited in the following way: they grew up their grove, but there were enough tangerines only for themselves—only this limited their work. And Tsvetkov, in a low bourgeois whisper, said, that he needed credit, that he needed orchard heaters. . . . I am certain that our Sochi region will be transformed into Sicily, into California, and we will destroy the mood, that is even now against citrus culture.[42]

Further criticism of the use of grove heaters emphasized their cost. As one critic wrote on the pages of *Sovetskie subtropiki* in 1934, "The method of experimental heating has led to many discussions. Those against the practice point to the great cost of the oil, the necessity of building large oil tanks, the difficulties of transporting, organizing and managing these heaters and doubts about the effectiveness of their use in the conditions of mountain relief."[43] Economic concerns deterred the development of the new technology.

Despite the open protest of Metelev, however, the Administration of Subtropical Agriculture chose to overlook the use of orchard heaters in Sochi. Despite the widespread and public celebrations of the success of Soviet breeding sciences in creating a more frost-resistant breed of citrus and the clear indication of the desires of the center for more laboratory research to reconstruct citrus from within, the use of orchard heaters quietly persisted. No conflict between center and periphery arose over orchard heating. Indeed, research into their use actually expanded, covered in the pages of the specialist journals, including *Sovetskie subtropiki*. Although economic concerns made adequate investments in agricultural technology impossible, there were limits to ideological attacks on their use.

Conclusion

Between 1928 and 1936, the Soviet state encouraged the development of subtropical agriculture in a newly defined economic region of the Soviet Union, the subtropics. The policy of the collectivization of agriculture included the promotion of monocultures on a large scale and the selection of optimal species for the broad dissemination in a given region. In the subtropics, these crops included citrus. The introduction of citriculture was framed as a challenge to Soviet agricultural science and practice, particularly as climatic obstacles threatened the new crop.

In the northernmost territory of the Soviet subtropics, the Sochi region, frequent frosts damaged the citrus groves and destroyed crops. During the particularly cold winter of 1928–1929, the Sochi Experimental Station experimented with the use of American-style orchard heaters and, finding their experiments successful, promoted their use within the region. The use of orchard heaters contradicted the central scientific policy that was emerging, which was focused on breeding new, frost-resistant species of citrus for the region. Particularly from 1935, Michurinism saw a more insistent rise in Soviet agricultural science and became state policy. In this context, an obscure breeder from the Sochi Experimental Station, F. M. Zorin, was elevated to the status of a national hero for his new frost-resistant breeds of tangerines. His work demonstrated the benefits of engaging Michurinist breeding methods.

In the spirit of the increasing political control of the agricultural sciences, local officials in Sochi denounced the use of the orchard heaters. Indeed, coverage of research into their use actually increased on the pages of the central organ of the central administration for subtropical agriculture, *Sovetskie subtropiki*. Instead, economic concerns initially deterred the development of the new technology.

As this chapter suggests, economic concerns formed the foundation of agricultural policy in the post-1931 period. Unable to sufficiently fund new technology at the local level, agricultural officials turned their attention to supporting approaches to agricultural problems that did not require large investments in machinery, particularly expensive machinery imported from abroad. Economic limitations were at the heart of the inability of local agricultural researchers to import technology from abroad, not an inherent belief in autarky. Policy makers chose to overcome these limitations as best they could by encouraging what seemed to be promising laboratory-based and practical research in breeding. At the local level, however, the economic foundation of the policy was inherently understood (as the citation of Boris Aznin suggested), if not directly articulated, and a relative amount of flexibility with experiments with technology was allowed in practice.

Notes

1. Karen Johnson Freeze, "Innovation and Technology Transfer during the Cold War: The Case of the Open-End Spinning Machine from Communist Czechoslovakia," *Technology and Culture* 48 (April 2007): 249–285.

2. György Péteri, "Nylon Curtain—Transnational and Transsystemic Tendencies in the Cultural Life of State-Socialist Russia and East-Central Europe," *Slavonica* 10 (November 2004): 113–123.

3. See K. E. Bailes, "The American Connection: Ideology and the Transfer of American Technology to the Soviet Union, 1917–1941," *Comparative Studies in Society and History* 23 (July 1981): 421–448. See also the articles in a special issue on technology of *The Russian Review*, "Technology: A Useful Category of Russian Historical Analysis," from July 2011, particularly the introduction by Scott W. Palmer, "Technology Defines Everything," *Russian Review* 70 (July 2011): 371–379.

4. On the use of foreign agriculture technology in the Soviet Union during the First Five-Year Plan, see Deborah Fitzgerald, "Blinded by Technology: American Agriculture in the Soviet Union, 1928–1932," *Agricultural History* 70, no. 3 (Summer 1996): 459–486, and Dana G. Dalrymple, "American Technology and Soviet Agricultural Development, 1924–1933," *Agricultural History* 40, no. 3 (July 1966): 187–206. Both sources emphasize the relatively low priority accorded to the acquisition and development of agricultural technology during those years of rapid industrialization, with the exception of the prized tractor. On the tractor, see Yves Cohen, "The Soviet Fordson: Between the Politics of Stalin and the Philosophy of Ford, 1924, 1932," in *Ford, 1903–2003: The European History*, ed. Hubert Bonin, Yannick Lung and Steven Tolliday (Paris: Éditions P.L.A.G.E., 2003), 2:531–558.

5. Warren R. Schoonover and Robert W. Hodgson, *Orchard Heating in California* (Berkeley: University of California Printing Office, 1925), 7.

6. Robert W. Hodgson Papers, 1918–1966, Finding Aid, University of California, Berkeley.

7. Frost Prevention Company, *The Bolton Orchard Heater, Known Also as the 'Fresno Pot' and the 'California Oil Pot': The Pioneer without a Peer* (San Francisco, 1911), 3.

8. Frost Prevention Company, *The Bolton Orchard Heater*, 3.

9. Schoonover, *Orchard Heating in California*, 28–29.

10. Schoonover, *Orchard Heating in California*, 32.

11. Schoonover, *Orchard Heating in California*, 29.

12. J. Anissimoff, *Soviet State Farms and Specialization in Agriculture* (Moscow: Lenin Academy of Agricultural Sciences, 1930).

13. Sovkhoz, "Iuzhnye kul'tury," *Krasnodarskii krai gor.* (Adler: Narkomzdrav SSSR Glav. Upr. Subtropicheskikh kul'tur, 1937).

14. Arkhivnyi otdel administratsii goroda-kurorta Sochi (Archive Department of the Administration of the City-Health Resort Sochi, AOAGKS), f. 29, op. 1, d. 2.

15. AOAGKS, f. 16, op. 1, d. 130, l. 129.

16. N. A. Baril'chenko, "Ekonomicheskaia kharakteristika raiona deiatel'nosti sochinskoi opytnoi stantsii," *Trudy Sochinskoi opytnoi stantsii subtropicheskikh I iuzhnykh plodovykh kul'tur* 9 (1935): 27.

17. The Sochi specialists wrote to a state farm in the region that they had not yet been able to import any "American" orchard heaters.

18. AOAGKS, f. 16, op. 1, d. 164, l. 17. Soviet experts established that tangerines were more frost resistant than oranges and therefore selected tangerines for cultivation in frost-prone regions.

19. Baril'chenko, "Ekonomicheskaia kharakteristika," 9, 12.

20. I found no evidence of the employment of the practice of smudging, coal-burning heaters, or other technologies of orchard heating prior to these experiments with the Bolton-type heaters in the Sochi archives.

21. Baril'chenko, "Ekonomicheskaia kharakteristika," 18.

22. N. K. Ianushevskii, "Vsesoiuznoe soveshchenie po zashchite subtropihcheskikh kul'tur ot moroza," *Sovetskie subtropiki,* 1–2 (1934): 253.

23. G. T. Selianinov, "O klimaticheski vozmozhnykh granitsakh kul'tury tsitursovykh v SSSR," *Sovetskie subtropiki* 1–2 (1934): 256–262, here 261.

24. The first name of Iakhontov was not given in the archival sources.

25. AOAGKS, f. 16, op. 1, d. 130, l. 12.

26. AOAGKS, f. 16, op. 1, d. 130, l. 63.

27. AOAGKS, f. 16, op. 1, d. 130, l. 63.

28. AOAGKS, f. 16, op. 1, d. 130, l. 11.

29. According to one account, Soviet purchases of agricultural equipment in the United States fell from $60,820,000 in 1930 to $3,498,000 in 1931 and dropped still further to $152,891 in 1932. See Dalrymple, "American Technology and Soviet Agricultural Development," 190.

30. See Bailes, "The American Connection," 423.

31. A. M. Lezhava, "Sozdadim sovetskuiu floridu," *Ogonek,* spetsial'nyi nomer (1934): 32–34.

32. S. Ashkhatsava, "Razvitie sotsialisticheskogo subtropicheskogo sel'skogo khoziaistva I zadachi nauchno-issledovatel'skoi raboty," *Sovetskie subtropiki,* 1–2 (1934): 3–5.

33. David Joravsky, *The Lysenko Affair* (Cambridge, MA: Harvard University Press, 1970), 77.

34. Joravsky, *The Lysenko Affair,* 53.

35. Joravsky, *The Lysenko Affair,* 73.

36. Ashkhatsava, "Razvitie sotsialisticheskogo subtropicheskogo sel'skogo khoziaistva," 12.

37. Ashkhatsava, "Razvitie sotsialisticheskogo subtropicheskogo sel'skogo khoziaistva," 13.

38. Ashkhatsava, "Razvitie sotsialisticheskogo subtropicheskogo sel'skogo khoziaistva," 5.

39. Ia. Usherenko, "Mandariny v Sochinskom raione," *Pravda,* 13 October 1935.

40. AOAGKS, f. 16, op. 1, d. 169, l. 49.

41. AOAGKS, f. 16, op. 1, d. 169, l. 49.

42. AOAGKS, f. 3, d. 48, l. 13.

43. V. P. Ekimov, "Otkrytoe oteplenie tsitrusovykh plantatsii, ego effektivnost' i perspektivy v SSSR," *Sovetskie subtropiki* 1–2 (1934): 272.

Bibliography

Anissimoff, J. *Soviet State Farms and Specialization in Agriculture.* Moscow: Lenin Academy of Agricultural Sciences, 1930.

Arkhivny otdel Administratsii goroda-kurorta Sochi (AOAGKS).

Ashkhatsava, S. "Razvitie sotsialisticheskogo subtropicheskogo sel'skogo khoziaistva i zadachi nauchno-issledovatel'skoi raboty." *Sovetskie subtropiki* 1–2 (1934): 3–5.

Baril'chenko, N. A. "Ekonomicheskaia kharakteristika raiona deiatel'nosti sochinskoi opytnoi stantsii." *Trudy Sochinskoi opytnoi stantsii subtropicheskikh i iuzhnykh plodovykh kul'tur* 9 (1935): 9–37.

Bailes, K. E. "The American Connection: Ideology and the Transfer of American Technology to the Soviet Union, 1917–1941." *Comparative Studies in Society and History* 23 (July 1981): 421–448.

Cohen, Yves. "The Soviet Fordson: Between the Politics of Stalin and the Philosophy of Ford, 1924, 1932." In *Ford, 1903–2003: The European History,* ed. Hubert Bonin, Yannick Lung, and Steven Tolliday, vol. 2, pp. 531–558. Paris: Éditions P.L.A.G.E., 2003.

Dalrymple, Dana G. "American Technology and Soviet Agricultural Development, 1924–1933." *Agricultural History* 40, no. 3 (July 1966): 187–206.

Ekimov, V. P. "Otkrytoe oteplenie tsitrusovykh plantatsii, ego effektivnost' i perspektivy v SSSR," Itogi rabot po promyshlennoi kul'ture tsitrusovykh v SSSR i perspektivy ee razvitiia." *Sovetskie subtropiki* 1–2 (1934): 271–274

Fitzgerald, Deborah. "Blinded by Technology: American Agriculture in the Soviet Union, 1928–1932." *Agricultural History* 70, no. 3 (Summer 1996): 459–486.

Freeze, Karen Johnson. "Innovation and Technology Transfer during the Cold War: The Case of the Open-End Spinning Machine from Communist Czechoslovakia." *Technology and Culture* 48 (April 2007): 249–285.

Frost Prevention Company. *The Bolton Orchard Heater, Known Also as the 'Fresno Pot' and the 'California Oil Pot': The Pioneer without a Peer.* San Francisco, 1911.

Graham, Loren. *What Have We Learned About Science and Technology from the Russian Experience?* Stanford, CA: Stanford University Press, 1998.

Hodgson, Robert W., Papers, 1918–1966. University of California, Berkeley.

Ianushevskii, N. K. "Vsesoiuznoe soveshchenie po zashchite subtropihcheskikh kul'tur ot moroza." *Sovetskie subtropiki* 1–2 (1934): 252–253.

Joravsky, David. *The Lysenko Affair.* Cambridge, MA: Harvard University Press, 1970.

Lezhava, A. M. "Sozdadim sovetskuiu floridu." *Ogonek,* spetsial'nyj nomer (1934): 32–34.

Palmer, Scott W. "Technology Defines Everything." *Russian Review* 70 (July 2011): 371–379.

Péteri, György. "Nylon Curtain—Transnational and Transsystemic Tendencies in the Cultural Life of State-Socialist Russia and East-Central Europe." *Slavonica* 10 (November 2004): 113–123.

Schoonover, Warren R., and Robert W. Hodgson. *Orchard Heating in California.* Berkeley: University of California Printing Office, 1925.

G. T. Selianinov, "O klimaticheski vozmozhnykh granitsakh kul'tury tsitursovykh v SSSR," *Sovetskie subtropiki* 1–2 (1934): 256-262.

Selianinov, G. T. "O klimaticheski tselesoobraznom razmeshchenii tsitrusovykh na territorii subtropicheskoi zony." *Sovetskie subtropiki* 1–2 (1934): 263–265.

CHAPTER 3

Swabian Water Treatment Technology in Russia

A Case Study of International Knowledge and Technology Transfer between West and East in the Late 1960s

Thomas Schuetz

Research Associate

Section for the History of the Impact of Technology (WGT)
History Department
University of Stuttgart
Stuttgart, Germany

The development of water treatment technologies is, with very few exceptions, still a desideratum in the history of technology and science, a fact that might be explained by its status as a fundamental technology that is frequently used but hardly ever acknowledged or celebrated. Because water treatment has been very closely linked to the corresponding scientific progress, changing views of customers and other relevant social groups, and the social and economic interaction, it is a promising field for research. In 2007, an in-depth study of the history of water treatment experts in the Stuttgart Metropolitan Area, which included a broad-based oral history project, was performed by the Historical Institute of the University of Stuttgart.[1] Some of the results of that research project are presented in the following. The case of the water treatment plant for the Togliatti car factory, located about 1,000 km southeast of Moscow in Samara Oblast at the Volga, is of exceptional interest for the issue of international technology transfer, with particular attention being paid to the movement of artifacts. In this case, a redundant ion exchange train was developed and built in Swabia, Germany, by the plant manufacturer Hager + Elsässer and was delivered to Togliatti in 1969. The aim was to recycle rinse water from electrogalvanization of car parts such as bumpers and headlights to reclaim the dissolved substances and reuse them for production. This knowledge and technology transfer in the middle of the Cold War[2] represents a typical phenomenon for the manufacturing industries of the Federal Republic of Germany after World War II, when technologically advanced products where developed and marketed for the international market in an environment with comparable high production costs. In the presented case the customer was the Soviet Union, which gives the case of Togliatti a special quality because it was a knowledge and technology transfer not just across borders but between the two competing political, cultural, and ideological systems of capitalism and communism.

Togliatti

In the Soviet Union Nikita Khrushchev and his administration had always preferred the expansion of public transport. The private ownership of a car seemed ideologically suspicious. Nevertheless, their seven-year plan of 1958 promised a massive increase in the national output of cars, with the aim of around 750,000 to 856,000. After Khrushchev's fall it became apparent that virtually no efforts had been made to reach these ambitious numbers and that the country would need international help to achieve the motorization of the nation. Thus, in 1966, the Central Committee decided that the citizens of the Soviet Union deserved the cheap, reliable, and practical car that had been promised to them eight years earlier. The fact that the Western "public enemies" had already had such vehicles, such as the Ford Model T, the Volkswagen Beetle, and the Citroën 2CV, for a considerable time turned the attempt to produce a comparable car into a question of national prestige.[3] Together with Fiat, where the Italian Communist Party played a dominant role, a large car manufacturing complex was established in the town of Stawropol-Wolschskij at the Volga.[4] The town was renamed Togliatti after the deceased leader of the Italian Communist Party, Palmiro Togliatti (1893–1964). The factory was designed to produce a single car model, a slightly modified version of the Fiat 124 named Schiguli (later better known as Lada), following the methods of Fordism and Taylorism. One small component of this large industrial complex was galvanization, and an even smaller part was the reclamation of the rinse waters from this shop. Before the late 1950s, these highly toxic wastewaters were normally discharged into the nearest river. Treatment of these wastewaters was considered only when economic interests were affected.[5] In the Soviet Union such environmental concerns were never raised. The fact that the Togliatti car factory was nevertheless equipped with the technology to clean the rinse water from the galvanization shop can be explained only by the fact that the Soviets wanted to establish a state-of-the-art factory that was equipped with all the features a modern industrial production facility could offer at that time.[6]

The contact between Hager + Elsässer and the Soviet Union was made via an established relationship with Blasberg, situated in Solingen, Germany, a well-known manufacturer of galvanization applications that had worked together with Hager + Elsässer for quite some time. Blasberg had virtually no water treatment expertise at that time and hired Hager + Elsässer as a subcontractor.[7] In 1967, the chief engineer of Hager + Elsässer, Kurt Marquardt (1930–2009), and the staff from Blasberg met with the responsible trade representatives of the Soviet Union in Moscow, and thanks to his capability as a salesman, he returned to Stuttgart with the biggest deal the company had ever brokered.[8] But what was the incentive for the Soviets to purchase this technology? To the surprise of the Germans, neither their economical nor their ecological arguments held sway; the emblematic character[9] of rinsing water through ion exchange as advanced and modern technology was sufficient to sell the technology to the officials of the Communist Party.[10]

Hager + Elsässer

Hager + Elsässer is a medium-sized Swabian plant manufacturer that had specialized in water treatment technology from its start in 1932. In the 1960s, Hager + Elsässer had production facilities in Stuttgart-Vaihingen, Holzgerlingen, and Sirchingen. In 1975, six years after the Togliatti deal, it had approximately 200 employees and annual sales of about 28 million Deutschmark. The treatment of rinse water resulting from electrogalvanization and recycling of its constituents for reuse in production had already been developed by Hager + Elsässer in 1959, and the company had previously gained experience in large-scale applications. These few core figures show that Hager + Elsässer is a typical example of a Western German small and medium enterprise (SME; *Mittelstandsunternehmen*) at that time.[11] To understand why such a company could participate in a globalizing economical world, it is necessary to have a brief look at the technology in question.

Ion Exchange Technology

The basic principle of ion exchange in water treatment had already been established for a considerable period of time. It is a technology based on the research of nineteenth-century British scientists John T. Way and Harris S. Thompson. Their observation that natural zeolites were capable of retaining ammonia and rejecting calcium led German chemist Leo Gans, after further research, to the conclusion that synthetic zeolites could be used as means of water purification.[12]

The first applications of this innovative technology were in the field of water treatment, where it was used in combination with different forms of the established chalk-natron technology.[13] The aim was to remove hardness components from feed water to protect steam engines and turbines and to increase the security of steam boilers. However, very quickly, breweries, the textile industry, and laundries used the technology as well because it enabled them to reduce their operating costs while preserving their machinery.

The ion exchange principle could not be patented, only the configuration of the ion exchangers. Such ion exchangers, first granulated material and later resins, were produced by the chemical industry,[14] whereas the applications to use them were developed by plant manufacturers. In the late 1950s and early 1960s, a small group of plant manufacturers specialized in the field of water treatment and the application of ion exchange technology. The geographical distribution of these experts in water cleansing showed a remarkable characteristic. Apart from Permutit in Berlin, Christ in Basle, and Degrémont in Paris, the majority were situated in Swabia:

Carl Morgenstern,[15] Streicher, Philipp Müller,[16] and W. Götzelmann + Partner in Stuttgart; EUWA in Gärtringen; Gütling in Fellbach; and, last but not least, Hager + Elsässer in Stuttgart-Vaihingen. Also, there were remarkable relations between the different companies even though they remained competitors. Hanns-Heinz Eumann, the founder of EUWA, was an apprentice at Hager + Elsässer and acquired Morgenstern in 1965.[17] Willy Hager (1905–1975), the founder of Hager + Elsässer, had worked for Carl Morgenstern before he established his own company in 1932. Finally, Hager + Elsässer acquired Streicher in 1978, Permutit in 1981, and Philipp Müller in 1998. The concentration of local experts in water treatment technology made it impossible for a company to survive if only a local or even national market would have been their sales territory. So the urge to go global can be explained on the basis of the regional particularities in the given example: a local agglomeration of competitors.[18]

In the late 1950s, these companies searched for new fields of application for their ion exchange technology. During this phase, the path to the development of large applications was not determined at all: industrial series production of rather small ion exchangers for small businesses and even households was a possible direction for future development.[19] For example, although it is little known, ion exchangers can be found today in such everyday household appliances as dishwashers. However, in the long run, this field of production did not yield the desired revenues for Hager + Elsässer. It turned out that their future remained in industrial applications and innovative technologies. The development of continuously working ion exchangers and the adoption of reverse osmosis technology proved to be crucial for the survival of this SME, situated in a technologically advanced region in a high-cost country.

The first tests for the detoxification of rinse water from electrogalvanization shops was carried out by the Hager + Elsässer engineers as early as 1959. The wastewater from industrial electroplating processes poisoned streams, waters supplies of other industries, and municipal sewage treatment plants. These wastes contained cyanide, chromate, acids, and alkaline solutions, as well as metals, such as copper, nickel, zinc, cadmium, chromium, iron, and others. These toxic substances can endanger the flora and fauna in lakes and streams, and they can also severely harm or even kill the microorganisms in biological wastewater treatment. They can affect the self-cleaning capacity of flowing waters and sludge decomposition in sewage treatment plants. It was inevitable that these toxic wastewaters would have to be treated. The contemporary conviction that it was sufficient to dilute the wastewater with fresh water had been rejected by the relevant professional associations and scientists as "irresponsible."

As early as 1960, Hager + Elsässer delivered an ion exchange water treatment recirculation system to IBM in Sindelfingen.[20] The first generations were still manually controlled, but in 1968 Hager + Elsässer developed its own control cabinet in Stuttgart-Vaihingen (Figure 1).[21] According to customer demands, after treatment in the ion exchanger train, the reclaimed material could either be reused in production or deposited as sewage sludge. In the case of electroplating, it was possible to keep freshwater consumption low and the resources in the production cycle. In 1966, the freshwater price alone made these applications a commercial success. The economy

1.
Ion exchange reactor from Hager + Elsässer on delivery to Hoechst in the early 1960s. Hager und Elsässer (Stuttgart-Vaihingen)/Wirtschaftsarchiv Baden-Württemberg (Hohenheim), Hoechst 01.

of such appliances depended significantly on the price of freshwater. In 1968, an American survey estimated that a freshwater price of 35 cents per 1,000 gallons would make an ion exchange train profitable, whereas in Germany, five years later the trigger price was 40 pfennige per cubic meter.[22] Therefore, for Western customers the reason to deploy an ion exchange water treatment system was basically an economical decision.

The Togliatti Project

The Togliatti project was not at all easy for Hager + Elsässer to implement. The reactors were much larger than anything Hager + Elsässer had ever produced. The application had a projected capacity throughput of 750,000 L/h.[23] The roof of the structure in Holzgerlingen was not high enough for the reactors, and the gantry crane was only capable of lifting a weight of 5 tons, whereas a single reactor for Togliatti weighed approximately 7 tons.

Funding was also problematic. Willy Hager, the founder and, in 1969, still director of Hager + Elsässer, had established his company at the culmination of the worldwide economic crisis in the 1930s, according to his self-created legend, to escape unemployment (Figure 2). Even though a profound analysis of the historical sources revealed that the founding of Hager + Elsässer was less dramatic than described in the retrospective, the experience of the crisis remained valid for the economic strategy of the founder. He had built up his company predominantly by reinvesting profits in the growth of the company and had always tried to avoid the influence of outside capital. However, the sheer size of the Togliatti project and comparable other contracts at the same time made it impossible to finance them out of the company's funds alone; outside money was needed. If such a major contract, which represented half the annual sales volume, had failed, the results would have been disastrous for Hager + Elsässer.[24]

2.
Kurt Marquardt (left) and Willy Hager. Hager und Elsässer (Stuttgart-Vaihingen)/Wirtschaftsarchiv Baden-Württemberg (Hohenheim), M17.

Since 1965, the percentage of major contracts in annual sales continuously grew, and Hager + Elsässer was forced to adapt established and innovative technologies in new environments as the size of the applications constantly became bigger. Against this background, the provisions for customer claims and for potential warranty exposure were increased, a reasonable reaction that further weakened the capital base of the company.[25] The decision to participate on a global market proved to be challenging in the case of Hager + Elsässer, and it might be suggested that this is a typical phenomenon for SMEs that decide to expand their sales territories.

Transfer

The actual production and transfer of the Togliatti complex remained a significant instant in the memory of all participants, although it is interesting that neither the archival material nor the interviews of those involved show any hints of challenges that arose from the installation and early operation of the plant as one might suspect. Water cleansing by ion exchange was an established technology at the time, but the mere size of the project was predominant in the reception. Apart from the upstream gravel and carbon filters the application consisted of eight large ion exchange reactor vessels, with a combined flow rate of 750,000 L/h.

3.
The reactors for Togliatti in front of the Hager + Elsässer workshop in Holzgerlingen. Hager und Elsässer (Stuttgart-Vaihingen)/Wirtschaftsarchiv Baden-Württemberg (Hohenheim), Togliatti 02.

The political setting of the Cold War[26] and the near past of the Second World War gave the interaction between Swabians and Russians a special quality. The Nazi era was still well remembered. Willy Hager's brother had died during the war in Russia, and Rudolf Orcellet, who was the only former Nazi party member on the board, had served as a guard for Russian POWs.[27]

The reactors and all large parts were transported from the production facilities in Holzgerlingen (Figure 3) by heavy goods trucks to the closest train station, where the large reactors where hauled on freight wagons. Surprisingly, in the commemoration of the contemporaries, this actual movement of objects remained insignificant. But what was remembered and also enhanced by anecdotes was the actual encounter between the Russians and Swabian workers that took place in Holzgerlingen, not in Togliattigrad. When consumables[28] for the Togliatti complex were fetched by a group of Russian laborers, who had traveled the whole 3,500 km from Togliatti at the Volga to Holzgerlingen by truck, the contemporary witnesses remembered that their only food consisted of cold borscht. I also found other stereotypes for poverty in the description of the Russians, such as worn-out shoes and used tires on their truck. Kurt Marquardt organized proper provisions at his own expense. A shared meal, accompanied by liquor, is said to have ended in a tear-soaked fraternization.[29] Apart from these rather ambivalent anecdotes, the sources remained almost completely silent about the actual transfer. Unfortunately, Kurt Marquardt, the decisive man behind the whole deal, who traveled to Togliatti to bring the application into service, could not participate in the oral history project. He suffered a stroke in 1992 and was an invalid without the ability to speak until his death in 2009. But thanks to the generous support of his family, his notebooks were available for the project. These notebooks contained at least hints of the problems that might have occurred. He had used simple sketches that might represent the applications, and these sketches were repeatedly overwritten. Such austere and rare relicts are as good as the narratives of contemporaries' quite ambivalent historical sources, which require careful interpretation and historical contextualization. The history of technology and economy—especially in the case of smaller and medium-sized enterprises, which do not maintain their own historical archives or museums—is always faced with this challenge.

Conclusion

An unexpected and telling result of the historical investigation was that the actual technological artifact, which had brought Russians and Swabians together, had almost disappeared from the memory of the contemporaries. The example of the technology transfer between Stuttgart and Togliatti shows that the motivation to participate in the global market for a SME like Hager + Elsässer can be dictated by its local environment and can be enabled by the technological gap between producer and customer, an observation that is confirmed by the different reception of technology in the East and the West during the period under investigation. However, on the way to a global market Hager + Elsässer was struggling with issues of scale: of the project, of its own company, of the client, and of the financing requirements. This struggle is likely a recurring issue for companies operating in a global market, but to what extent Hager + Elsässer and the water purification by ion exchange can be seen as a paradigmatic example for the participation of a SME in a global market or whether it was a special case remains a desideratum for further research.

Notes

1. Thomas Schuetz and David Seyffer, *Hager + Elsässer: Die ersten 75 Jahre—Von Menschen und Leidenschaft* (Stuttgart: Hager + Elsässer, 2007). See also Historical Institute of the University of Stuttgart, Unternehmensgeschichte, http://www.uni-stuttgart .de/hi/gnt/forsch/forschung11.htm (accessed 15 March 2012).

2. Yale Richmond, *Cultural Exchange and the Cold War: Raising the Iron Curtain* (Philadelphia: Pennsylvania University Press, 2000), xiv.

3. Lewis H. Siegelbaum, *Cars for Comrades: The Life of the Soviet Automobile* (New York: Cornell University Press, 2008), 80–124.

4. "SOWJET-UNION / FIAT-FABRIK Eintreffe morgen" (editorial), *Der Spiegel*, 9 (26 February 1968), 100.

5. "Vaihinger Firma leistet Pionierarbeit" (editorial), *Stuttgarter Zeitung*, 20 January 1971.

6. Rolf Nagel (head of chemical engineering, Hager + Elsässer), in discussion with the author, 9 July 2006.

7. Kurt Waldbauer (technical section leader, retired, Hager + Elsässer), interview by the author, 28 November 2006.

8. Notebook of Kurt Marquardt (CEO and former CTO, Hager + Elsässer), 4 October 1967, private collection.

9. Ulrich Wengenroth, "Gute Gründe: Technisierung und Konsumentenentscheidung," *Technikgeschichte* 71, no. 1 (2004): 3–18. See Uwe Schimak, "Rationalitätsfiktion in der Entscheidungsgesellschaft," in *Zur Kritik der Wissensgesellschaft*, ed. Dirk Tänzler (Konstanz, Germany: UVK Verlagsgesellschaft, 2006), 57–81.

10. Nagel, discussion with author, 9 July 2006.

11. Bernd Venohr and Klaus E. Meyer, *The German Miracle Keeps Running: How Germany's Hidden Champions Stay ahead in Global Economy* (Working Paper 30, Fachhochschule für Wirtschaft, Berlin, 2007), 3.

12. Permutit Aktiengesellschaft, *50 Jahre Permutit Aktiengesellschaft: 50 Jahre im Dienst des Ionenaustausches* (Berlin: Permutit AG, 1962), 5.

13. David Seyffer, "Innovationsforschung in der Unternehmensgeschichtsschreibung: Beispiele aus der Historie der Firma HAGER + ELSÄSSER," in *Wissenschaft und Technik als Motoren unternehmerischen Handelns*, ed. Thomas Schuetz and David Seyffer (Stuttgart: GNT- Verlag, 2007), 65–100.

14. Erik Verg, Gottfried Plumpe, and Heinz Schultheis, *Meilensteine: 125 Jahre Bayer 1863–1988* (Köln: Informedia-Verlag, 1988), 420.

15. Hanns-Heinz Eumann (apprentice at Hager + Elsässer and founder of EUWA), in discussion with the author, 27 March 2007.

16. Ulrike Zubal and Gerhard F. Volkmer, *90 Jahre Philipp Müller: Eine Chronik; Wasseraufbereitung 1896–1986* (Stuttgart: Philipp Müller Nachf. Eugen Bucher, 1986).

17. EUWA H. H. Eumann GmbH, Unternehmen–History, http://www.euwa.com/index.php/de/unternehmen/historie (accessed 9 January 2016).

18. Bruce Mazlish, *The New Global History* (New York: Routledge, 2006), 66.

19. Hannelore Brändle (former secretary and later cohabitee of Willy Hager, founder of Hager + Elsässer), interview by the author, 1 May 2007.

20. Hager + Elsässer, *50 Jahre 1932–1982* (Stuttgart: Hager + Elsässer, 1982), 16.

21. Nagel, discussion with author, 9 July 2006.

22. Robert Weiner, *Die Abwässer der Galvanotechnik und Metallindustrie*, 4th ed. (Bad Saulgau, Germany: Leuze, 1973), 53.

23. "Von Stuttgart an die Wolga" (editorial), *Stuttgarter Zeitung*, 11 February 1969.

24. Horst Geidel (CEO, Behr; son-in-law of Willy Hager), interview by the author, 2 June 2007.

25. Hager + Elsässer, closing balance sheet, 1965 and 1966, Wirtschaftsarchiv Baden-Württemberg, University of Hohenheim, Stuttgart.

26. Rana Mitterer and Patrick Major, *Across the Blocs: Cold War Cultural and Social History* (London: Routledge, 2004), 2.

27. Rudolf Orcellet (technical director, then advertising manager, retired, Hager + Elsässer), interview by the author, 2 November 2006.

28. "Rohrzeugs und so spezielle Sachen haben die mitgenommen," Reinhold Arndt (foreman of the welding shop, retired, Hager + Elsässer), interview by the author, 14 December 2006.

29. Arndt, interview by the author, 14 December 2006.

Bibliography

Hager + Elsässer. *50 Jahre 1932–1982*. Stuttgart: Hager + Elsässer, 1982.

Mazlish, Bruce. *The New Global History*. New York: Routledge, 2006.

Mitterer, Rana, and Patrick Major. *Across the Blocs: Cold War Cultural and Social History*. London: Routledge, 2004.

Permutit Aktiengesellschaft. *50 Jahre Permutit Aktiengesellschaft: 50 Jahre im Dienst des Ionenaustausches*. Berlin: Permutit AG, 1962.

———. *Permutit Taschenbuch*. 6th ed. Berlin: Permutit AG, 1953.

Richmond, Yale. *Cultural Exchange and the Cold War: Raising the Iron Curtain*. Philadelphia: Pennsylvania University Press, 2000.

Schimak, Uwe. "Rationalitätsfiktion in der Entscheidungsgesellschaft." In *Zur Kritik der Wissensgesellschaft*, ed. Dirk Tänzler, pp. 57–81. Konstanz, Germany: UVK Verlagsgesellschaft, 2006.

Schuetz, Thomas, and David Seyffer. *Hager + Elsässer: Die ersten 75 Jahre—Von Menschen und Leidenschaft*. Stuttgart: Hager + Elsässer, 2007.

Seyffer, David. "Innovationsforschung in der Unternehmensgeschichtsschreibung: Beispiele aus der Historie der Firma HAGER + ELSÄSSER." In *Wissenschaft und Technik als Motoren unternehmerischen Handelns*, ed. Thomas Schuetz and David Seyffer, pp. 65–100. Stuttgart: GNT-Verlag, 2007.

Siegelbaum, Lewis H. *Cars for Comrades: The Life of the Soviet Automobile*. New York: Cornell University Press, 2008.

Venohr, Bernd, and Klaus E. Meyer. *The German Miracle Keeps Running: How Germany's Hidden Champions Stay Ahead in Global Economy*. Working Paper 30, Fachhochschule für Wirtschaft Berlin, Berlin, 2007.

Verg, Erik, Gottfried Plumpe, and Heinz Schultheis. *Meilensteine: 125 Jahre Bayer 1863–1988*. Cologne: Informedia-Verlag, 1988.

Weiner, Robert. *Die Abwässer der Galvanotechnik und Metallindustrie*. 4th ed. Bad Saulgau, Germany: Leuze, 1973.

Wengenroth, Ulrich. "Gute Gründe: Technisierung und Konsumentenentscheidung." *Technikgeschichte* 71, no. 1 (2004): 3–18.

Wirtschaftsarchiv Baden-Württemberg. University of Hohenheim, Stuttgart.

Zubal, Ulrike, and Gerhard F. Volkmer. *90 Jahre Philipp Müller: Eine Chronik; Wasseraufbereitung 1896–1986*. Stuttgart: Philipp Müller Nachf. Eugen Bucher, 1986.

CHAPTER 4

Canada, Communism, and the Colombo Plan

David McGee

Archivist

Canada Science and Technology Museums Corporation Ottawa, Ontario, Canada

Rian Manson

Independent Scholar

Ottawa, Ontario, Canada

In the archives of the Canada Science and Technology Museum are three large rolls of linen paper, held together by wooden slats. They look rather like ancient scrolls (Figure 1). They are, in fact, a set of arrangement drawings for the last steam locomotives ever built in Canada. Delivered in 1955 and 1956, these locomotives were part of the largest order ever received by the Canadian Locomotive Company (CLC) of Kingston, Ontario. The 120 WP 4-6-2 Pacific-type locomotives completed by CLC were not, however, built for use in Canada.[1] They were built for India under the Colombo Plan for Co-operative Economic Development in South and Southeast Asia. Still in existence, but now mostly forgotten, the Colombo Plan was the first organized attempt to enroll South and Southeast Asia in the Cold War project of Western globalization. Canada played an important role in this project.

Although the concept is greatly contested, it can be argued that there are two basic schools of thought about globalization.[2] The first sees the phenomenon in terms of the "annihilation of distance" resulting from new technologies that make contact and communication between distant regions ever more rapid. Such, for example, were the Western technologies of shipbuilding, navigation, and telegraph that bound India into the British Empire. The story of the Canadian locomotives built for India under the Colombo plan is not a story of this type. Although they might be considered agents of communication, the locomotives delivered by CLC were already obsolete in North America at the time they were ordered. Moreover, they were built as part of

1.
Three rolls of drawings for the 120 Indian WP class 4-6-2 Pacifics built by the Canadian Locomotive Company in 1955 and 1956. CSTMC/STR Collection: Image CLC001.

a plan to make India self-sufficient in rail production, rather than enrolled in global patterns of trade and industry.[3]

A second school of thought sees globalization in terms of social relations, particularly the development of a greater social interdependence and interconnectedness between regions that were once separated by geographical distance.[4] Acknowledging preexisting connections, such as those of India's incorporation into the British Empire, this school sees the dramatic increase in the intensity and fixity of global social interactions as the product of Cold War pressures leading to the increased participation of separate regions in the operations of international capitalism and then to global sharing, or even homogenization, of social and cultural values. Frequently, this second school of thought sees globalization in terms of large-scale historical forces leading to unintended consequences as, for example, in the frequently described growth of multinational corporations leading to the diminishing power of the nation-state.[5]

The story of the Canadian locomotives built under the Colombo Plan is definitely a tale of this second kind, but in many ways a corrective one. The result was definitely an increase in the intensity, frequency, and fixity of contact between Canada and India, which, apart from their membership in the British Commonwealth, had almost no social or economic relations prior to World War II. The increased postwar contact was also a direct product of the Cold War, in the sense that the Colombo Plan was originally conceived as a way to prevent the international spread of communism. Nevertheless, this early step in the history of globalization was not a matter of global forces leading to unintended consequence for the nation-state. Rather, it was a deliberate negotiation designed to strengthen a nation-state, namely India. Furthermore, although these negotiations were indeed based on a shared language of economics and diplomacy, what is most striking is not the social values shared by Canada and India but the asymmetry of both language and expectations. On one side, the Indian government set out to improve the Indian economy, armed with the language of classical economics and the expectation of an eventual improvement in the Indian standard of living. On the other, the Canadians felt themselves bereft of economic theory as it applied to "underdeveloped" countries and were extremely skeptical about the likely results of Canadian aid with respect to improvement in Indian living conditions.

Where do steam locomotives fit into this picture? If the story of globalization is one of increasing frequency, fixity, and intensity of social relations, the answer is that it was the locomotives provided by CLC that cemented the new global relationship between Canada and India, functioning very much as "actants" in a hybrid network as described in the actor-network theory of John Law and Bruno Latour.[6] Until the provision of locomotives as aid was agreed to, Colombo Plan relations between Canada and India were floundering. Once agreed to, the pattern of the relationship stabilized and became fixed, and the scale of the interaction increased rapidly. This was initially a small initial step in the history of postwar globalization, but it was a step that had a major effect on global history—for it was on the pattern of the WP locomotives provided under the Colombo Plan that Canada also provided India with a nuclear reactor. It was with that reactor that India constructed its nuclear bomb.

The Origins of the Colombo Plan

To understand the genesis of the Colombo Plan in 1950, it will be useful to begin with a brief outline of the most important political and economic development of the previous decade, particularly as they affected South and Southeast Asia.

The place to begin is with the fact that both Canada and the United States emerged from World War II with thriving economies and a globalizing agenda. Focusing on Europe, this agenda aimed at rebuilding the international financial system through the creation of various new institutions and programs. The Marshall Plan is the most famous of these initiatives, but both Canada and the United States also made large direct loans to Britain in order to shore up the British pound. The most menacing challenge to the Canadian and American agenda was the continuing advance of world communism, which saw the installation of Soviet-backed regimes in several Eastern Bloc countries immediately after World War II, followed by a brutal Soviet coup in

Czechoslovakia, the blockade of Berlin, the formation of the German Democratic Republic (East Germany), and the "fall" of China to communism in 1949.[7]

The economies of the various nations in South and Southeast Asia were in an even more dreadful state than those of Europe, having never recovered from the Great Depression before they were further ravaged during World War II. The economic state of both India and Pakistan were further stressed by the dissolution of the British Empire, which led to independence in 1947 but also to partition and war, resulting in the displacement of millions of refugees whose needs had to be taken care of in the midst of a food crisis.[8] A potential bright spot for India and Pakistan was the large "sterling balances" that both countries had built up in London during World War II, which were credits for the supply of military goods and services that Britain was unable to pay for at the time. These large balances, however, led to serious macroeconomic problems for Britain and its former colonies, the central problem being that Britain's postwar economic problems left it still unable to pay its debts or export manufactured goods that it could sell to the colonies for sterling. Instead, Britain and its former colonies were forced to purchase both capital and consumer goods from the "dollar area," meaning Canada and the United States, for the most part. Purchasing such goods meant, in effect, selling sterling for dollars. If not carefully managed, the result could be a sudden devaluation of the currencies of all sterling denominated currencies. This is precisely what happened in 1947 and in 1949, when "sterling crises" led to runs on the pound, threatening to destroy the economies of India and Pakistan as well as Britain.[9]

The sterling balances were of great concern to Canada and the United States because of the loans both countries had made to Britain just after the war. The purpose of the loans was to restore British productive capacity as one step in a larger plan to restore multilateral trade around the world. The repeated sterling crises threatened to undo what Canada and the United States had been able to accomplish so far. From the point of view of globalization, the sterling balances created a financial link between Canada and India, despite the fact that Canada had nothing to do with their creation and essentially no commercial or trading relationships with that country.

As former colonies, Canada and India were linked by social and political relations of a different kind. They were both members of the former British Commonwealth, an organization threatened with collapse when India and Pakistan gained their independence in 1947 and promptly went to war with each other. Not until 1949 was this crisis resolved, with Canada playing an important role by helping to broker the deal that allowed India and Pakistan to remain in a newly constituted "Commonwealth of Nations."[10] The survival of the Commonwealth was an important step in the history of postwar globalization because of its function as an international organization that linked widely separated regions of the world. It provided a forum for discussion of international issues at its annual meetings and a mechanism with which problems could be resolved on the basis of shared history, values, and language. The survival of the Commonwealth is also important because without it there could have been no Colombo Plan. The recently weathered crisis of the Commonwealth ensured that discussion of the initiative took place in an atmosphere of utmost respect for the new sovereignty and dignity of countries like India and Pakistan and an awareness of the potential fragility of relations between Commonwealth countries. This awareness had a major impact on the way the Colombo Plan was organized.

These elements of the political and economic situation form the essential background to the Colombo Plan, so called because the first of the three meetings that led to the creation of the plan took place in Colombo, Sri Lanka, in January of 1950. In deference to the newly independent former colonies of India, Pakistan, and Ceylon (now Sri Lanka), it was the first meeting of the Commonwealth to take place in Asia. Its main purpose was to provide the foreign affairs ministers of the Commonwealth countries with an opportunity to discuss the world political situation, particularly the Asian situation in light of the communist takeover in China in 1949. However, a second meeting took place at the same time, at which more junior officials were charged with finding a permanent solution for the problem of the sterling balances.[11] In other words, there was initially no "Colombo Plan" on the agenda in Colombo. Nevertheless, an appeal was made by Ceylon for direct economic aid to the area at the political meetings of foreign ministers and was supported by Australia, India, and Pakistan. The appeal led to joint discussions between the senior foreign affairs ministers and the junior economic officials. The result was an agreement to form a Commonwealth Consultative Committee. This committee would meet in Sydney in May to examine the possibilities.[12]

The Sydney meeting proved to be a stormy one, largely because the Australians demanded the immediate commitment of large amounts of emergency aid to the countries of the region, a demand that was strenuously resisted by Britain, New Zealand, and Canada, whose officials had explicit instructions not to agree to anything of the sort. India, Pakistan, and Ceylon supported the Australian position at first but eventually agreed with the British and Canadian counter-argument, which was that the amount of aid actually available from Commonwealth countries was not enough to bring about any significant economic improvement in South and Southeast Asia. For that they needed American dollars. If they hoped to attract the Americans (incidentally easing the problem of the sterling balances by bringing large amounts of American dollars into the sterling area), they would need to create a practical, carefully thought-out plan.[13] It was therefore agreed that a third meeting would take place in London in September, giving the recipient countries time to make a thorough study of their economic situation. At the London meeting, the Consultative Committee presented a detailed report on the causes of the economic problems faced by the countries of South and Southeast Asia and set out a long list of projects for which potential recipient countries could use external aid. The report was approved, and the various delegations returned home to recommend to their governments that the Colombo Plan for Co-operative Economic Development in South and Southeast Asia be supported.[14]

It is relatively easy to demonstrate that one of the main reasons for Canada's support of the Colombo Plan was the need to fight the advance of communism in Asia. From Colombo, for example, Canadian External Affairs Minister (later Prime Minister) Lester Pearson reported his belief that through the proposed Consultative Committee "a great deal may be done not only to solve the problem of the sterling balances but also to shore up our defences in this area against the tide of Soviet expansionism."[15] On his return home, Pearson told his fellow cabinet ministers that "the programme would be designed both to strengthen the economies of the countries in the area and to help combat the spread of communism."[16] Instructions to the Canadian delegation attending the second meeting in Sydney were also explicit:

> The Delegation should express the concern of the Canadian Government over conditions in South and Southeast Asia. The Government is aware of the urgency of at least making a start in improving the standard of living in such countries as India, Pakistan, Ceylon, Burma, Malaya, Indo China, Indonesia and Thailand if the spread of Communism is to be prevented.[17]

Fighting the spread of communism was on everyone's mind at the London meeting because the Korean War had just started. Canadian delegates were instructed that "the military aggression against South Korea in no way diminished but, on the contrary, accentuated the need for improved economic, political and social conditions in Asia."[18] The delegates reported to Ottawa the general feeling at the London meeting that "the West must take whatever steps were open to it to prevent any further large segments of the Eurasian land-mass from falling under Communist domination."[19]

These developments clearly support theoretical arguments concerning the role of the Cold War in promoting post-World War II globalization. Whereas, prior to 1939, the Canadian government had no perceived interest and no sense of responsibility for living conditions in India or Pakistan, the need to fight communism changed that, creating conditions in which Canada was suddenly interested in committing to a plan to improve the standard of living on the other side of the earth. It is worth stressing the point made by the senior Canadian economic adviser to the three Colombo plan meetings, Douglas LePan, who has written that the establishment of the Colombo Plan marked the first time that the needs of any of the *underdeveloped* nations of the world had ever received comprehensive, detailed attention, and it was the first time that large amounts of financial aid to *developing* countries had ever been considered.[20]

How was financial aid going to help defeat communism? The general idea was that aid would help raise the standard of living of people in countries like India and Pakistan, who would therefore have no need choose communism out of desperation. How was the aid going to raise the standard of living? The answer to this question was more complicated. According to LePan, because the question of financial aid to underdeveloped countries had never been considered, the Canadians had no economic theory to guide them.[21] This may have been true, but the real stumbling block Canadian officials faced in answering this question was their own deep skepticism about the likely results of Canadian aid to impoverished countries like India and Pakistan. This skepticism had already shown itself at the Sydney meetings, where it was argued that not even the total amount of aid available from Commonwealth countries would make a difference and only the Americans had the kind of money needed. Such skepticism may also be seen in the instructions to the Canadian delegation, who were ordered to resist all attempts to simplify the problem of raising the standard of living in the area by treating it as a purely economic matter and to stress that Canada could not even consider the question of economic aid until basic social, cultural, and even corruption elements of the problem had been carefully examined. As the official instruction continued,

> The attainment of higher standards of living and development in South and Southeast Asia must inevitably depend very largely upon the efforts of the peoples and governments themselves. The role of outside assistance can, at most, be one which is directed to the provision of "missing components" which may be most helpful in the carrying out of comprehensive domestic programmes.[22]

In short, the Canadian position was that the people of the region would really have to do it themselves, which would take a long time. In this context, aid could only help make a start by providing at most missing components in the form of technologies that countries like India and Pakistan could not supply for themselves.[23]

If, however, financial aid was going to have so little effect on the standard of living in the short term, how was it going to have any effect on the pressing need to fight communism in the long term? Here, the Canadians were forced into the position that aid would be *symbolic*, signaling to the people and especially the elites of the region that the West was willing to help. It was difficulties that the Canadians had in creating the appropriate symbolic regime that led to the building of the Indian WPs. Before discussing these difficulties, it will be important to look at the Indian situation in greater detail.

India and the Origin of the WP Pacifics

Canadian officials may have felt a lack of appropriate economic theory, but Indian officials had no such problem. Committed to state planning of the economy even before independence, they had done a great deal of formal thinking about what to do and why, drawing on classic capitalist, Keynesian, and Marxist theory. Their thinking was not only expressed in policy documents but embedded into India's formal five year plans, the first of which began in 1951. These plans naturally placed a great deal of emphasis on agriculture, considering the country was overwhelmingly agrarian yet faced with constant food shortages, but the five-year plans also stressed the production of capital goods and therefore the development of heavy industry as the only way to sustain a permanent increase in the standard of living over time.[24] It was further argued that since only the state had the capital needed to create the necessary industry, direct state ownership of different industries was required, especially the railways.[25]

Railways were of crucial importance to India, given that the country possessed neither an extensive system of roads nor an extensive system of navigable waterways. Indeed, at the time of independence, India had one of the largest rail networks in the world. The state of the network was, however, deplorable. One set of problems stemmed from the fact that prior to World War II, most of the lines were owned and operated by individual British interests, which led to the establishment of literally hundreds of different locomotive classes and the purchase of almost all locomotives from Britain, leaving India with virtually no indigenous building capacity. Another set of problems was the result of World War II, when thousands of miles of track, millions of sleepers, and 30% of all locomotives were relocated either to the Middle East or Burma.[26] Almost no reinvestment in new track or rolling stock took place, leaving postwar India with a system consisting of miles of worn-out track and worn-out rolling stock. Since the capital investment required to build was massive, the existing railway lines were nationalized, then combined into a single administrative system.

Nationalization was accompanied by two further decisions. One was to develop an indigenous ability to construct locomotives and thereby build up at least some of the heavy engineering capacity thought to be needed for long-term economic growth. The second was to standardize railway operations wherever possible, including the design of passenger locomotives.[27] The Indian Railways therefore prepared its own specification, then turned to Baldwin Locomotive Works in the United States to complete the design and produce the initial engines. Thus was born the Indian WP class of 1947, where the *W* stood for the wide-gauge Indian track (5 feet, 6 inches), and *P* stood for passenger locomotive.

The decision to build steam rather than diesel locomotives is not perfectly understood. One factor was certainly the fact that India had large supplies of coal but little money to buy foreign oil. Another factor was the idea that steam locomotives were a kind of heavy engineering that India's railways could learn to build quickly. A third may have been the fact that Indian railways had no prior experience with diesels and thus no way to devise a standard diesel locomotive adapted to local conditions, whereas they had a great deal of experience with steam and had already conducted a great deal of research into the kind of standard locomotive needed.[28] The decision to work with Baldwin is easier to explain. Near the end of World War II, the Indian railways were in such bad shape that the only way to keep them running until the struggle with Japan was over was to import hundreds of North American–built 2-8-2 Mikados. Indian officials were impressed with these rugged machines, discovering that many features of American practice were well adapted to the Indian context. A large number of these wartime locomotives were built by Baldwin, which was arguably the world's leading builder of steam locomotives in 1947.

The design completed by Baldwin was based on the 4-6-2 Pacific type, of which many thousands had been built in North America between 1900 and 1945, but nevertheless blended an array of British, American, and Indian features. For example, the WPs had double roofs and high windows that were specifically designed to help crews cope with the tropical Indian heat. They had large American-style "Boxpok" driving wheels, as well as American-style grates and fireboxes, which were able to burn the low-quality, high-ash Indian coal. The normal Pacific-type wheel placement was changed to reduce axle loadings and thereby save wear and tear on the fragile Indian track. British practice was followed with regard to the vacuum and steam brakes, leading bogies, valve gear, rings, linkage gear, and other details, but American practice was followed in making parts that were interchangeable within the WP class and even interchangeable

2.
Indian WP Pacific in the erecting shops of the Canadian Locomotive Company in 1949. The locomotive in the air reflects the streamlining of the WPs. Below is a standardized WP boiler on its frame. CSTMC/STR Collection: Image CSTM STR 34748.

between different classes. For example, the WP boilers, tenders, axle boxes, springs, boiler mountings, valve gears, cabs, and several other components were all designed to be interchangeable with India's new 2-8-2 WG class of freight trains (with the *G* standing for "goods").[29] To sum up, although steam was regarded as obsolete in North America in the late 1940s and early 1950s, steam locomotive engineering actually reached its peak of perfection just after 1945. By turning to Baldwin and borrowing best practices from around the world, India was able to use the most modern steam locomotive technology available in 1947. That they could do so points to a perhaps surprising level of globalization with respect to the steam engineering of the time.

One further borrowing deserves comment, which is the striking bullet-nosed "streamlining" of the WPs, which drew on the German immigrant Otto Kuhler's outstanding design for the Baltimore and Ohio's Royal Blue locomotive of 1937, which was, incidentally, a 2-6-2 Pacific (Figure 2). The streamlined style as adopted by such design luminaries as Raymond Loewy and Henry Dreyfuss was applied to such famous trains as the New York Central Railway's Twentieth Century Limited. Having more or less no effect on performance, streamlining was purely symbolic, sending a message of modernization and progress.[30] The style was extremely popular in the

1930s but, like steam technology in general, increasingly obsolete in North America after 1945. It was not obsolete in India, however, where the bullet-nose streamlining of the WPs was deliberately chosen for its symbolic value, a point made perfectly clear by Indian Ambassador to the United States M. Asaf Ali, who, when taking ceremonial delivery of the first WP from Baldwin in 1947, stated explicitly that the WPs "were streamlined as a psychological factor in publicizing the idea of modernization."[31]

Engines of Stabilization

Given India's continuing need for locomotives and Canada's need to provide financial aid to South and Southeast Asia, it might seem that orders for the WP locomotives from CLC should have been made under the Colombo Plan in 1951. Instead, it took two years for a particular set of problems with the Colombo Plan to become clear to the Canadians. These problems had to do with symbol management. Until they were solved, the new global relationship between Canada and India remained unstable. The WPs were the solution.

A major source of problems was a long delay in getting the Colombo Plan off the ground. One cause of delay was the reluctance of many Canadian cabinet ministers to accept the recommendation of the Canadian delegation to the final Colombo Plan meeting in London and to agree to fund the program. The idea was very popular with the public, but several members of the cabinet believed the country could not afford the expense, given its other international financial commitments to Europe, Britain, the United Nations and NATO. Others argued that the best way to fight communism was to rearm, not waste money on the hopeless economies of India and Pakistan. It took almost five months for External Affairs Minister Pearson to overcome these objections. Not until 21 February 1951 was he able to announce that Canada would participate in the Colombo Plan. The level of spending was set at $25 million for the first year, $15 million of which was to go to India and $10 million to Pakistan.[32]

A second source of delay was continued wrangling over how the Colombo Plan, as what would now be called an international government organization (or IGO), would be organized. Would there be a large or small bureaucracy? A permanent bureaucracy or not? And with what responsibilities? Negotiations involving all the Commonwealth countries took all of 1951 and continued into 1952 before it was decided that the secretariat would be very small and its duties confined to the organization of technical and educational exchanges and to the gathering of data about possible projects to be funded.[33] The secretariat was not empowered to decide *which* projects received funding or to disburse cash since that would appear as if the former imperialist masters, posing as donors, were still controlling the economies of their former colonies. In deference to the sensitivities of the newly independent colonies, it was agreed that all aid would be worked out on a bilateral basis between donor and recipient countries directly.

A third issue to be settled was how aid money was to be managed and spent. On this topic the Canadian government insisted on the use of what were called "counterpart funds," meaning funds established in local currency that were equivalent (or "counterpart") to the dollar value of the donation and that would be spent on projects agreed to by the donor.[34] Thus, in the first year of the Colombo Plan, Canada agreed to donate $10 million in wheat to alleviate the food crisis India was experiencing at the time. India sold the wheat in local markets around the country, using the proceeds to set up a counterpart fund in rupees. The rupees were then used to help finance the Mayurakshi hydroelectric and irrigation project. Negotiating these financial arrangements took time. India only agreed in June of 1951 and Pakistan in August of that year.

The protracted negotiations helped to create the first of the symbolic problems Canadian officials faced with respect to the Colombo Plan. The public had been told that both the humanitarian and anticommunist need was pressing. The cabinet had been repeatedly told that the best way to fight communism in Asia was make a start on improving the economies of countries like India and Pakistan. Parliament had approved $25 million. Yet, at the end of the first year of the Colombo Plan, 1951–1952, less than half the $25 million had been spent. India got its $10 million in wheat but no more, and Pakistan received no aid at all. It therefore looked like the Canadian government did not really care about conditions in South and Southeast Asia and had not only duped its own people but broken its symbolic promise to the desperate people of South and Southeast Asia. This was not the sort of signal that would convince the people of the region to reject communism, nor was it the kind of issue the government wanted to debate in the House of Commons, which would have to be asked to approve the carryover of unspent money from one year to the next. The cabinet therefore approved a rather extraordinary measure under which the unspent Colombo money was formally "given" to Indian and Pakistan but actually deposited with the Canadian Commercial Corporation (CCC), a Crown corporation originally set up by the government in 1946 to facilitate Canadian exports to Europe through the negotiation of government-to-government contracts. India and Pakistan agreed to designate CCC as their official agent.[35]

The CCC manoeuver eliminated the need for a debate in Parliament but did not solve the main problem. In the second year of the Colombo Plan, 1952–1953, Canada pledged another $25 million and again failed to spend the money.[36] Only $5 million in wheat went to India. Some aid made its way to Pakistan. In total, according to reports, only $18 million of the $50 million approved in the two previous fiscal years had been spent. That left $32 million—more than half.[37]

A major reason for the lack of spending was disagreement with India over what Canada was willing to provide under the Colombo Plan. In accordance with the "missing components" theory, officials expected to provide India with money that India would use to acquire equipment. India wanted more wheat. Canadian officials explained that wheat was one of Canada's most important cash exports and the country could not afford to keep giving it away.[38] Indian officials explained that they preferred wheat that they could sell to create counterpart funds because they could use counterpart funds in a flexible manner. The Canadians expressed their surprise that India was not interested in the Western technology they clearly needed. The Indians replied that they were not able to order Canadian equipment because India had already embarked on its first five-year plan, under which all orders for capital equipment had already been placed. Adding to Canadian surprise, Indian officials went on to state that Canadian equipment purchased with counterpart funds was generally more expensive than the same equipment they could buy on the world market. This meant that when the cost of the equipment was charged to a particular project, it was more than the project managers had budgeted for, upset the financing for that project, and was therefore actually giving Canada a black eye.[39]

From the perspective of time, it seems almost humorous. Here was Canada offering millions and India refusing to take it. But Indian officials were, in fact, explaining the essence of a second symbolic problem that Canadian officials were also beginning to recognize. As one Canadian diplomat reported,

> I was told in the South that they would welcome aid from us (and God knows they need it, particularly in Madras now in its sixth year of drought) but they hated getting all tangled up in the red tape of the Central Government. In any case, the Central Government makes them pay counter-part funds to the full extent of every nut and bolt, and so they were much happier to get a grant from the Central Government, or to use their own Provincial Funds and buy where they liked, and above all, have complete control over deliveries, co-ordination, etc., etc., which enabled them really to get something done. You will remember the Central Government had almost to beat West Bengal into taking $3 million from us for Mayurakshi electrical equipment. It is the same story in every State: Canadian aid is just a bothersome, restricting and very expensive business to them. It is no wonder that Canada makes no friends from her aid programme.[40]

In short, Canada was not getting credit for what it was doing and therefore not earning the symbolic gratitude required if communism was to be defeated.

Canadian officials responded to this challenge by developing an even stronger desire to deliver capital goods and equipment to India under the Colombo Plan, in accordance with an explicit theory about what they called the "psychological advantage" of doing so. What "psychological advantage" meant was the unavoidable reality of an actual object right in front of a person clearly identified as coming from Canada, making it impossible for that person not to recognize the source and presumably feel grateful, something that was not happening with purchases made in dribs and drabs from counterpart funds. As one internal memo stated in September of 1952, "a contribution in wheat, even though it produces badly needed counterpart funds for use by the Indian Government, lacks the psychological (and possibly commercial) advantages which the provision of identifiably Canadian equipment might have."[41] Exactly the same arguments were made at the cabinet level.[42]

Consideration then turned to the question of what kind of equipment would fit the symbolic bill. The answer was the provision of railway equipment on a large scale, which would have all the psychological advantage one could ask for *and* make it possible to spend all the money voted by Parliament for the Colombo Plan, thereby solving both problems at the same time.

Delivering the Order

India appears to have agreed to the delivery of railway equipment once convinced that the flow of wheat would stop and possibly in recognition of the fact that wrangling over wheat had already cost them $15 million dollars in potential aid over two years. Things then proceeded rather quickly. On 26 March 1953, the Canadian government approved the expenditure of $2.2 million for the construction of forty locomotive boilers from the Montreal Locomotive Works for delivery to India, later adding ten more boilers to the order.[43] On 9 September 1953, the cabinet approved $11 million for the construction of sixty to sixty-five steam locomotives to be built by CLC.[44]

CLC was at that time Canada's oldest locomotive manufacturer, having been in business for more than 100 years. The company thus had a staff with an immense amount of experience in building steam locomotives. It also had experience building locomotives for India, delivering two hundred sixty 2-8-2 Mikados between 1943 and 1950. It even had prior experience with WPs, delivering eighty WPs to India in 1949–1950.[45] One complication was the fact that CLC had been acquired in 1950 by Fairbanks-Morse of the United States, which set out to modernize CLC's product line with an opposed-piston diesel unit for the Canadian market. To make room for diesel production, CLC eliminated much of its heavy steam locomotive engineering capacity, including its boiler shop. This change resulted in higher than expected unit costs for the WPs, which led India to order 120 locomotives in an effort to bring unit costs down. The cabinet approved this doubling of the order on 29 December 1953 at an additional cost of $10 million. Five million in wheat was reluctantly also approved, which meant the government was suddenly able to spend two years of Parliamentary appropriations in advance.

Production of the Indian WPs began as soon as CLC signed a contract with the Canadian Commercial Corporation in March 1954. The first locomotive was shipped from Kingston in March of 1955 (Figure 3). The remainder were completed at the rather astonishing pace of one locomotive every four days.[46] The last WP was completed in September of 1956—the date that the last steam locomotive built in Canada was shipped to India for reassembly by Indian workers (Figure 4).

In India, the WPs were a great success. A total of 755 WPs from all sources were constructed between 1947 and 1967, making them one of the most successful locomotive classes of all time. Of the total, 259 were built in the Chittaranjan Locomotive Works, where India succeeded in building up the heavy engineering capacity called for under its five-year plans.[47] Individually, the WPs proved to be such good locomotives that the last of them was only withdrawn from service in 1996. That was a quarter century longer than CLC, which went out of business in 1969. The WPs also seem to have been successful messengers of modernism, becoming symbols of Indian pride in the 1960s and 1970s when they were used to haul such prestigious passenger trains as the Taj Express, the Grand Trunk Express, the Howrah-Madras Mail, and other "superfasts" (Figure 5).

The WP program was a success for Canada too. It allowed the government to spend all the money appropriated by Parliament for the Colombo Plan, thereby sending the message to India

3.
Dignitaries in Kingston, Ontario, in 1955, attending the delivery of the first of the 120 WPs built by CLC for India under the Colombo Plan. President of the Canadian Commercial Corporation William Low is third from left. High Commissioner for India M. A. Rauf is fifth from the right. Several other CLC and government officials are in the picture. CSTMC/STR Collection: Image CSTM STR 29834.

4.
The last steam locomotive built in Canada was tested in the CLC yards before it was knocked down into sections and loaded on flatcars for transport in 1956. CSTMC/STR Collection: Image CSTM STR 25439.

5.
Canadian built WP #7615 in the Saharanpur locomotive depot in 17 February 1992, after many years of service. Many of the modernist bullet noses of the WPs were painted in bright colors with the Star of India on the front. Photo by Roger Morris - Buriton Wheelbarrow Rail Photos.

that Canada did care about conditions in that country, and it allowed officials to send the kind of capital equipment for which they felt Canada would earn the gratitude of individual Indians whenever they rode the train. To what extent Canada actually earned credit from the WPs is probably impossible to say, but Canadian officials seemed satisfied, which allowed the new global aid relationship between India and Canada to settle into a regular pattern of interaction.

Conclusion

We started this chapter by describing the Colombo Plan as an early step in postwar globalization, taking globalization to be a process of constructing new social relations between geographically distant parts of the world. We have focused on the construction of new relations between Canada and India, showing that the process had little to do with the rise of multinationals, the declining power of nation-states, or unintended consequences brought about by historical forces. On the contrary, it was a deliberate creation of nation-states that was intended to strengthen one of them, namely, India, in the hope that it would strengthen India's resolve in the fight against communism.

The Colombo Plan certainly led to closer ties between the two countries, including forms of social, political, and economic interaction that simply did not exist prior to World War II. It was not inevitable that the new relationship should become fixed. A number of specific practices had to be created, ranging from Colombo Plan studies through Parliamentary appropriations, contracts with Crown Corporations, the creation and use of counterpart funds, and so on. Even so, two years after the creation of the Colombo Plan, the relationship between Canada and India (at least from the Canadian point of view) was floundering. The heart of the problem was Canadian skepticism with respect to the likelihood that economic aid to India would have any practical effect in the short run. This skepticism led Canadian officials to focus on the symbolic nature of aid and to worry that Canada was not getting the credit for its symbolic gestures.

Skepticism, however, should not be confused with cynicism. As senior Canadian economic advisor Douglas LePan observed years later, historians might dismiss the Colombo Plan as just another Western ploy in the Cold War battle against communism or charge that Canada participated in the plan only to help its own economy or argue that the whole plan was really a form of covert Americanism, designed to bring U.S. influence via U.S. dollars into the sterling area. Against these charges, LePan replied that one of the most powerful arguments made against the Colombo Plan in the cabinet was that the best way to fight communism was to rearm. If it was a matter of Canadian self-interest in boosting its own economy, naysayers would have to explain the fact that the biggest opponent of Canadian participation in the Colombo Plan was the minister of finance. Those who thought it was all about getting American dollars would have to reckon with the fact that Canada announced its support for the Colombo Plan in Parliament before knowing what the Americans would do.[48] LePan's major point was that numerous contingent factors had to be overcome in order to build a stable relationship. This required willpower and the ultimate source of that willpower was a genuine desire to help alleviate poverty in South and Southeast Asia. "We believed," he wrote, "that on a radically shrunken and shrivelled planet, the peoples of the earth could not permanently endure, half rich and half poor. Just like those who thought that there had to be a bridging of the gap between rich and poor inside a country, there was a movement of opinion towards the idea that efforts had to be made to narrow the gap between rich and poor countries."[49] Here is a statement of the globalization of consciousness if there ever was one.

What is interesting from the theoretical point of view is that Canadian officials recognized that they needed a certain kind of technology to stabilize the network of relations they were building, particularly that they needed capital equipment—big stuff. The WP locomotives fit the bill admirably, helping to settle Canada's new global relations with India into a pattern. Once the locomotives were delivered, however, the pattern required more big stuff. Canadian documents show that even before the last of the WPs was delivered, India and Canada were negotiating something bigger still. This was the provision of India's first nuclear power plant, which went online in 1960. This provision of nuclear technology again allowed Canada to spend and even increase the aid given to India under the Colombo Plan and entailed even closer sociotechnical relations. It was also the reactor India used to build its first atomic bomb, adding a whole new dimension to Cold War globalization.

Notes

1. The *W* stands for wide gauge, and the *P* stands for passenger, as designated by the Indian Railways. These locomotives had four pilot wheels, six driving wheels, and two trailing wheels and were thus 4-6-2's according to the standard Whyte notation. There is no general agreement as to the origins of the "Pacific" designation; some believe the name arose because the first

two locomotives of this type were ordered by New Zealand railways, and others believe it is because the type was heavily used by the Missouri Pacific in the United States.

2. See the discussion of various views in Bruce Mazlish, *The New Global History* (Routledge: New York, 2006).

3. For a good overview see Albert J. Churella, *From Steam to Diesel: Managerial Customs and Organizational Capabilities in the Twentieth Century American Locomotive Industry* (Princeton, NJ: Princeton University Press, 1998).

4. See, for example, Jürgen Osterhammel and Niels P. Petersson, *Globalization: A Short History* (Princeton, NJ: Princeton University Press, 2003); Karl Moore and David Lewis, *The Origins of Globalization* (New York: Routledge, 2009); Jacques B. Gelinas, *Juggernaut Politics: Understanding Predatory Globalization* (New York: Zed Books, 2003); Eugene D. Jaffe, *Globalization and Development* (Philadelphia: Chelsea House Publishers, 2006).

5. Mazlish, *New Global History*, 32–33.

6. For descriptions of actor-network theory as it might apply to globalization, see Bruno Latour, *Science in Action: How to Follow Scientists and Engineers through Society* (Cambridge, MA: Harvard University Press, 1987); and Bruno Latour, *We Have Never Been Modern* (Cambridge, MA: Harvard University Press, 1993). See also John Law, "On the Methods of Long-Distance Control: Vessels, Navigation and the Portuguese Route to India," in *Power, Action and Belief*, ed. John Law, Sociological Review Monograph 32 (London: The Sociological Review, 1986), 235–263; and John Law, "Technology and Heterogeneous Engineering: The Case of Portuguese Expansion," in *The Social Construction of Technological Systems: New Directions in the Sociology and History of Technology*, ed. Wiebe Bijker, Thomas P. Hughes, and Trevor J. Pinch (Cambridge, MA: MIT Press, 1987), 111–134.

7. For an overview of these developments, see the various essays in Melvyn P. Leffler and Odd Arne Westad, eds., *The Cambridge History of the Cold War*, vol. 1, *Origins* (Cambridge: Cambridge University Press, 2010).

8. Brian Roger, "The State and the Economy of Modern India, 1939–1970: The Emergence of Economic Management in India," in *The Economy of Modern India, 1860–1970*, ed. Gordon Johnson, C. A. Bayly, and John F. Richards, The New Cambridge History of India 3.3 (Cambridge: Cambridge University Press, 1993), 160–165. See also A. Vaidyanathan, "The Indian Economy Since Independence (1947–1970)," in *The Cambridge Economic History of India*, vol. 2, *c. 1957–c. 1970*, ed. Dharma Kumar, Meghnad Desai, and Tapan Raychaudhuri (Cambridge: Cambridge University Press, 1989), 947–995.

9. Douglas LePan, *Bright Glass of Memory: A Set of Four Memoirs* (Toronto: McGraw-Hill Ryerson, 1979), 156–157.

10. See the relevant volumes of *Documents on Canadian External Relations*, an online collection provided by the Canadian Department of Foreign Affairs and International Trade (DFAIT), located at http://www.international.gc.ca/history-histoire/documents -documents.aspx (accessed 15 August 2013). This chapter relies heavily on the repository, but the documents are something of a nightmare to cite, having long and numerous headers, subheaders, multiple titles, multiple dates, and multiple official numbers. All the documents cited in this article are from the annual chapters on Commonwealth Relations. They will be identified as DFAIT. The proper volume and document numbers, date, and a representative title are provided, which should enable scholars to find the citation online, as well as in the original paper volumes.

11. LePan, *Bright Glass of Memory*, 157.

12. LePan, *Bright Glass of Memory*, 99, 173–181. For the official account of the meeting, see DFAIT, vol. 12, 652–654, 21 January 1950, *Meeting of the Commonwealth Foreign Ministers, January 21st, 1950*.

13. LePan, *Bright Glass of Memory*, 195–196. See also DFAIT, vol. 16, 663, 11 May 1950, *Meeting of Commonwealth Consultative Committee for Southeast Asia, Sydney*.

14. For the final report, see the Colombo Plan Consultative Committee, *The Colombo Plan for Co-operative Economic Development in South and Southeast Asia: Report by the Commonwealth Consultative Committee* (London: His Majesty's Stationary Office, 1951).

15. DFAIT, vol. 16-654, 21 January 1950, *Meeting of the Commonwealth Foreign Ministers*.

16. DFAIT, vol. 16-659, 2 May 1950, *Meeting of the Consultative Committee for South and Southeast Asia, Canadian Participation*.

17. DFAIT, vol. 16-659, 2 May 1950, *Instructions for the Canadian Delegation to the Meeting of the Commonwealth Consultative Committee on South and Southeast Asia, to be held at Sydney, Australia, the 15th of May, 1950*.

18. DFAIT, vol. 16-675, 11 September 1950, *Meeting of the Commonwealth Consultative Committee for Southeast Asia, September 25–October 4*.

19. DFAIT, vol. 16-682, 18 October 1950, *Interdepartmental Committee on External Trade, Policy Document*.

20. LePan, *Bright Glass of Memory*, 149.

21. It is worth remembering that the Marshall Plan of 1948 marked the first time in history that massive economic aid had ever been freely given by one country to another. That was money given to developed Western nations to help them get back to their previous level of prosperity. Giving large amounts of money in aid to underdeveloped countries was something brand-new.

22. DFAIT, vol. 16-659, 2 May 1950, *Instructions for the Canadian Delegation to the Meeting of the Commonwealth Consultative Committee on South and Southeast Asia, to be held at Sydney, Australia, the 15th of May, 1950*.

23. For further examples of pessimism about the results of aid, see DFAIT, vol. 16-685, 24 October 1950, *Memorandum to Cabinet*, where it is argued that "in countries like India and Pakistan, however, rapid progress is impossible; indeed the countries themselves do not expect quick results. Whatever progress they make will be slow and painful." See also DFAIT, vol. 17-543, 17 January 1951, *Secretary of State for External Affairs to Minister of Finance*, where Lester Pearson writes, "It is true, of course, that even if the $3 billion can be provided from external sources and if the programme is implemented in substantially the shape that is now proposed, there will not be any dramatic improvement in standards of living in South and Southeast Asia. There will be, for example, an increase of only some 10% in the volume of food grains produced. That is certainly modest enough when the present poverty of the area and the likely increase in population are taken into account."

24. Amal Sanjal, "The Curious Case of the Bombay Plan," *Contemporary Issues and Ideas in Social Sciences* 6 (2010). http://journal .ciiss.net/index.php/ciiss/article/view/78/75 (accessed 15 August 2013).

25. Tomlinson, "The State and the Economy of Modern India," 166–180; Vaidyanathan, "Indian Economy Since Independence," 949–957.

26. Jogendra Nath Sahni, *Indian Railways: One Hundred Years* (New Delhi: Ministry of Railways [Railway Board], 1953), 150.

27. Sahni, *Indian Railways*, 98.

28. Sahni, *Indian Railways*, 98–99. For a technical description of the CLC WPs, see "First of 120 Steam Locomotives for India," *Canadian Transportation*, 58 (1955): 169–170.

29. Sahni, *Indian Railways*, 99.

30. Otto Kuhler, *My Iron Journey: An Autobiography of a Life with Steam and Steel* (Denver: Intermountain Chapter National Railway Historical Society, Boulder, 1967); Robert Selph Henry and Otto Kuhler, *Portraits of the Iron Horse: The American Locomotive in Pictures and Story* (Chicago: Rand McNally, 1937); Herbert Harwood, *Royal Blue Line: The Classic B&O Train between Washington and New York* (Baltimore: Johns Hopkins University Press, 2002).

31. "Baldwin Delivers Locomotive to Indian Government," *Railway Age* 123 (16 August 1947): 295.

32. The struggle is described in LePan, *Bright Glass of Memory*, 218–222. For official documents reflecting opposition to the Colombo Plan, see DFAIT, vol. 16-689, 4 October 1950, *Meeting of Commonwealth Consultative Committee for Southeast Asia, London*; vol. 17-543, 17 January 1951, *Consultative Committee Meeting, February 12–20, 1951, Secretary of State for External Affairs to Minister of Finance*; vol. 17-546, 30 January 1951, *Consultative Committee Meeting, February 12–20, 1951, Minister of Finance to Secretary of State for External Affair*; vol. 17-548, 6 February 1951, *Consultative Committee Meeting, February 12–20, 1951, Extract from Cabinet Conclusions.*

33. See, for example, DFAIT, vol. 17-550, February 12–20, 1951 *Consultative Committee Meeting.*

34. DFAIT, vol. 17-563, 23 February 1951, *Aid to India, Pakistan and Ceylon, Minutes of Interdepartmental Meeting*; vol. 17-578, 23 July 1951, *Commonwealth Prime Ministers' Meeting, London, January 4–12, 1951, Colombo Plan Discussions Held in Ottawa June 21st to 28th*; vol. 17-580, 17 August 1951, *Commonwealth Prime Ministers' Meeting, London, January 4–12, 1951, Colombo Plan–Pakistan.*

35. DFAIT, vol. 18-610, 20 March 1952, *Colombo Plan, Arrangements for Carry-Over of Funds*; vol. 18-612, 11 August 1952, *Colombo Plan*; vol. 18-614, 2 September 1952, *Colombo Plan.*

36. DFAIT, vol. 18-650, 14 November 1952, *Colombo Group, Extract from Minutes of Meeting.* "It was reported that apart from the $5 million grant of wheat no funds have been spent or committed for projects in India during the current year."

37. Officials tried to argue that the money had been "committed," even if it had not actually been spent. DFAIT, vol. 20-397, 13 October 1954, *Next Year's Colombo Plan Contribution.*

38. Discussion related to wheat can be followed in a series of documents, including DFAIT, vols. 18-612, 18-640–18-652. See especially DFAIT, vol.19-621, 18 June 1953, *Colombo Plan Assistance to India.*

39. DFAIT, vol.18-651, 8 December 1952, *Record of Discussions with Mr. Bhattacharyya and Canadian Officials*; vol. 19-621, 18 June 1953, *Colombo Plan Assistance to India.*

40. DFAIT, vol.19-608, 27 March 1953, *Colombo Plan, Extract from Letter from Administrator of Colombo Plan to Director, International Economic Relations Division, Department of Finance.*

41. DFAIT, vol. 18-645, 8 September 1952, *Colombo Plan; Memorandum for Under-Secretary of Stale for External Affairs to Secretary of State for External Affairs.*

42. DFAIT, vol. 18-647, 13 September 1952, *Colombo Plan Wheat for India, Extract from Cabinet Conclusion.*

43. DFAIT, vol. 19-620, 26 March 1953, *Colombo Plan: Extract from Cabinet Conclusions*; vol. 19-623, 19 November 1953, *Colombo Plan, Additional Locomotive Boilers for India.*

44. DFAIT, vol. 19-622, 9 September 1953, *Colombo Plan; Programmes for India And Pakistan; Consultative Committee Meetings; Fourth Canadian Contribution.*

45. Don McQueen and William Thompson, *Constructed in Kingston: A History of the Canadian Locomotive Company, 1854 to 1968* (Kingston: Canadian Railroad Historical Association, Kingston Division, 2000), 66.

46. The calculation is given in McQueen and Thomson, *Constructed in Kingston*, 291.

47. Sahni, *Indian Railways*, 105–110.

48. LePan, *Bright Glass of Memory*, 223–224.

49. LePan, *Bright Glass of Memory*, 223.

Bibliography

"Baldwin Delivers Locomotive to Indian Government." *Railway Age* 123 (16 August 1947): 295.

Canadian Department of Foreign Affairs and International Trade. *Documents on Canadian External Relations*. http://www.inter national.gc.ca/history-histoire/documents-documents.aspx (accessed 15 August 2013).

Churella, Albert J. *From Steam to Diesel: Managerial Customs and Organizational Capabilities in the Twentieth Century American Locomotive Industry.* Princeton, NJ: Princeton University Press, 1998.

Colombo Plan Consultative Committee. *The Colombo Plan for Co-operative Economic Development in South and Southeast Asia: Report by the Commonwealth Consultative Committee.* London: His Majesty's Stationary Office, 1951.

"First of 120 Steam Locomotives for India." *Canadian Transportation* 58 (1955): 169–170.

Gelinas, Jacques B. *Juggernaut Politics: Understanding Predatory Globalization.* New York: Zed Books, 2003.

Harwood, Herbert. *Royal Blue Line: The Classic B&O Train between Washington and New York.* Baltimore: Johns Hopkins University Press, 2002.

Henry, Robert Selph, and Otto Kuhler. *Portraits of the Iron Horse: The American Locomotive in Pictures and Story.* Chicago: Rand McNally, 1937.

Jaffe, Eugene D. *Globalization and Development.* Philadelphia: Chelsea House Publishers, 2006.

Kuhler, Otto. *My Iron Journey: An Autobiography of a Life with Steam and Steel.* Denver: Intermountain Chapter National Railway Historical Society, Boulder, 1967.

Latour, Bruno. *Science in Action: How to Follow Scientists and Engineers through Society.* Cambridge, MA: Harvard University Press, 1987.

———. *We Have Never Been Modern.* Cambridge, MA: Harvard University Press, 1993.

Law, John. "On the Methods of Long-Distance Control: Vessels, Navigation and the Portuguese Route to India." In *Power, Action and Belief*, ed. John Law, pp. 235–263. Sociological Review Monograph 32. London: The Sociological Review, 1986.

———. "Technology and Heterogeneous Engineering: The Case of Portuguese Expansion." In *The Social Construction of Technological Systems: New Directions in the Sociology and History of Technology*, ed. Wiebe Bijker, Thomas P. Hughes, and Trevor J. Pinch, pp. 111–134. Cambridge, MA: MIT Press, 1987.

Leffler, Melvyn P., and Odd Arne Westad, eds. *The Cambridge History of the Cold War.* Volume 1, *Origins.* Cambridge: Cambridge University Press, 2010.

LePan, Douglas. *Bright Glass of Memory: A Set of Four Memoirs.* Toronto: McGraw-Hill Ryerson, 1979.

Mazlish, Bruce. *The New Global History.* Routledge: New York, 2006.

McQueen, Don, and William Thompson. *Constructed in Kingston: A History of the Canadian Locomotive Company, 1854 to 1968.* Kingston: Canadian Railroad Historical Association, Kingston Division, 2000.

Moore, Karl, and David Lewis. *The Origins of Globalization.* New York: Routledge, 2009.

Osterhammel, Jürgen, and Niels P. Petersson. *Globalization: A Short History.* Princeton, NJ: Princeton University Press, 2003.

Sahni, Jogendra. *Indian Railways: One Hundred Years.* New Delhi: Ministry of Railways (Railway Board), 1953.

Sanjal, Amal. "The Curious Case of the Bombay Plan." *Contemporary Issues and Ideas in Social Sciences* 6 (2010). http://journal.ciiss.net/index.php/ciiss/article/view/78/75 (accessed 15 August 2013).

Tomlinson, Brian R. "The State and the Economy of Modern India, 1939–1970: The Emergence of Economic Management in India." In *The Economy of Modern India, 1860–1970*, ed. Gordon Johnson, C. A. Bayly, and John F. Richards, pp. 160–165. The New Cambridge History of India 3.3. Cambridge: Cambridge University Press, 1993.

Vaidyanathan, A. "The Indian Economy Since Independence (1947–1970)." In *Cambridge Economic History of India.* Volume 2, *c. 1957–c. 1970*, ed. Dharma Kumar, Meghnad Desai, and Tapan Raychaudhuri, pp. 947–995. Cambridge: Cambridge University Press, 1989.

CHAPTER 5

Inkonvensional Pathways
Soldered Supply Chains from Indonesia's Tin Islands

Matthew Hockenberry
Media Historian

Department of Media, Culture, and Communication
New York University
New York, New York

The head of the village said the area where Rosnan was working was an illegal mine, and that it would be refilled and re-seeded. Yet even as Rosnan's family was mourning, three teenage boys, soaking in the rain, scraped for tin ore at the bottom of the same pit, right near the sapling . . . They knew about Rosnan's death, and kept digging. Rosnan worked among thousands of Indonesians who wield pickaxes and buckets each day on Bangka Island, extracting the tin that becomes the solder that binds components in the world's tablet computers, smartphones, and other electronics.

—Cam Simpson, "The Deadly Tin
Inside Your Smartphone," 2012.

Off of the eastern coast of Sumatra, between the South China and Java Seas, is the island of Bangka. A province of Indonesia, Bangka is home to over half a million people. The Dutch naturalist François Valentyn found it unremarkable. "It extends," he wrote in 1724, "from 2 to 3 degrees; there it forms a small but well-known Straits." The inhabitants (of which he recorded "about a thousand") had seen their numbers diminish over the previous century. Of the material foundations of their society, "they have wax, honey, iron, and cotton; they also have many foodstuffs."[1] Valentyn's exploration of the Indies was some years prior to the publication of his history and, consequently, is notable in lacking a key component of Bangka's future. There was no tin on Valentyn's Bangka. By the time his account was published, it had become the most significant product of the island's shores, the only promise for its future prosperity.[2]

The contemporary narrative linking the "deadly tin inside our smartphones" to the interior of this otherwise unfamiliar tropical island marks only one of the many routes joining the tiny pathways of consumer electronics to the global networks from which they have been assembled.[3] For most of the last century Bangka has appeared in these sorts of accounts only a handful of times.[4] But as concerned observers rediscover connections to Bangka's remote landscape, they redirect questions that once focused on its processing power to the troubling qualities of the signals passing through its interior. Bangka's legacy in the history of global production becomes reimagined as a deviant phenomenon or historic relic—a poorly fashioned conduit in an otherwise finely wrought network of production.

The global landscapes of contemporary production appear and disappear at the whim of the seemingly impersonal economic forces that shape them. For every one that emerges and each that fades away, others remain, marked by earlier machineries of fabrication, by legacies of distant trade, by colonial habitations. Global supply chains depend on these ancient networks of extraction, where individual inhabitants persist in pursuing their own uncertain hopes for the future. Beyond the limits of technical rationality, the movements of commodities and populations play out, in the language of Arjun Appadurai, amid diverse geographies of imagination, of habitation, of capital, and of culture.[5] These global commodity circuits feed back onto themselves, into their own histories, as local actors rehabilitate and reconfigure old flows toward new ends. The myriad "scapes" cutting across the scarred physical landscape of Bangka testify to this, a case study of how the global pathways of assembly—that is, of supply chains—are tied to the local configurations of people and materials. In this way one of the quintessential metals of modernity rests on the simultaneously careful and chaotic construction of the geographies of extraction on a small Indonesian island. Bangka becomes a site of disjuncture. Rather than the global construction of the local, we see also its reverse, where the legacy of colonial geographies collides with the failures of postcolonial and corporate rationalization.

Nearly 70% of the tin circulating throughout the world is mined in Southeast Asia's tin belt. Half comes from Indonesia, 90% of it supplied by Bangka and the neighboring island of Belitung. Much is mined outside of the direct control of industry or government. Surveys conducted in 2001 showed that so-called illegal mining units were responsible for about 42,000 tons of tin sand—more than the 40,000 tons produced "legally" by the Indonesian state's mining concern, PT Timah Tbk, and a figure accounting for roughly 20% of worldwide tin production for that year.[6] Bangka has become a crucial site for PT Timah's professional mining operations in the decades following the rise of the Suharto regime (Figure 1), but despite efforts at regulation by the Indonesian government, desperate bodies continue to be drawn to the porous pits scattered over its tin sands. In the language of global capitalism, Bangka's illegal miners hope to "develop their capacity," "realize their economic potential," and "overcome the challenges" borne from their otherwise limited access to resources.[7]

1.
"Kepindahan kantor pusat ke Bangka adalah hal yang mendesak" (Moving the head office to Bangka is a matter of urgency).
Cover of *Stannia*, PT Timah industry magazine, 1991.

Their efforts go by many names. Labor organizations refer to "small-scale mining," and the United Nations refers to "artisanal mining practice." But for PT Timah and industry groups, it remains simply "illegal."[8] None of these labels are descriptive of actual operations, which are not necessarily illegal or (as numbers suggest) even small scale. Illegality would suggest that a framework of fixed laws and rights govern Bangka, but arrests remain rare and unlikely to be brought to trial.[9] Despite increasing regulation, there is nothing akin to a comprehensive law governing tin extraction. Mining requires permits (without them, they are PETI—*Penambang Tanpa Izin*), but these remain, at best, a reflection of the extragovernmental structures built long in the island's past. In the broadest sense, Article 33 of the Indonesian Constitution seems to clearly outline the state's overriding right of control (*Hak Menguasai Negara*), permitting it to "exploit all the lands, waters, and natural riches contained therein for the greatest benefit of the people." But throughout the island's history nearly all levels of government have been able to issue mining permits, and they continue to do so—producing ambiguity that ensures miners traverse bifurcating networks of regulation along their pathways to opportunity. Conflicts are common, with an overlapping impression of authority that allows as much as 90% of mining to be regarded by the government as PETI. Facing prosecution or arrest, most operations are still able to produce local permits or point to past privileges to offer some justification for their claims.[10]

Regardless of the external qualifier—illegal, small scale, or artisanal—the kind of mining practices implicated by these labels are recognized in Indonesia as *inkonvensional* (unconventional), a term that avoids connotations of authorization and scale but is not without a misleading character. The proliferation of distributed mining operations on Bangka and Belitung is *not* the exception, and contrary to PT Timah's accounts of the island's industrial rationalization, they are not recent.[11] The routes by which global landscapes are connected to Bangka are strikingly circuitous ones. That is to say, they are indirect, but also they now begin with the circuit—the tiny pathways that have become necessary for modern social, cultural, and economic life. These pathways, I would suggest, are visible here not as the product of some abstractly global modernity but as the deeply embedded outcome of an awkwardly fashioned network of local manufacture, trade, and supply. A contemporary critique of deviant globalizations can trace symptoms to century-old regimes of extractive labor, the promise of flexibility, and the forcible binding of a local productive community. *Inkonvension*, in the case of Bangka, has been a persistent modality of mineral extraction and electronics production from the first telegraphs and telephones through to the consumer products of high technology and homogenized design. Tin and the tin solder that joins the transistors and chips in the circuits of electronic devices (Figure 2) bridge a sprawling supply chain fashioned from this historic circumstance. Outside the regimes of any particular government, organizational authority, or global ideology, the tin sands have produced a cadence and convention of their own—a convention that reaches back to the beginnings of the island's history and that has long seemed certain to continue into its future.

Sinking Ship

Ancient accounts tell that the twin islands of Bangka and Belitung were created when a sinking ship broke in two, the distinctive mountains of the islands forming from the masts of the mythic vessel (Figure 3). The name "Bangka" is sometimes said to derive from the Sanskrit *vanga* (*vangka* or *wangka*), meaning tin or lead.[12] It offers a metonymic origin, the word for its most important commodity having come to stand in for the island itself. Still, I find notable Karl Helbig's suggestion

2.
Tin soldering. Photograph, Carlin Wing.

3.
Jacques Nicolas Bellin, "Carte des Isles de Java, Sumatra, Borneo & les Detroits de la Sonde Malaca et Banca Golphe de Siam . . ." (ca. 1750). Map of Java and Sumatra, showing Bangka (Banca) and an unnamed Belitung.

4.
Tin workers on Bangka ca.1930. Tropenmuseum, part of the National Museum of World Cultures, Amsterdam, the Netherlands (http://www.tropenmuseum.nl).

that settlers on the island found inspiration in the corpses (*bangkai*) of previous inhabitants who had starved from the island's meager provisions. These imagined encounters reinforce the early reality of a tenuous landscape made habitable only through phenomenal struggle, a fragmentary fabric of trade and exchange. Without streams of tin under the hot island sands, it is difficult to imagine what lives could have been made on the island or who would have found any reason to travel there.[13]

The discovery of tin in the early 1700s brought with it a perpetual source of immigrants, both Chinese laborers and European traders (Figure 4).[14] The Dutch East India Company (VOC) had assumed authority in nearby Melaka following their victory over the Portuguese in 1641, and they entered the tin market through Bangka under the auspices of the Sultanate of Palembang in 1717. Their influence would persist, to varying degrees, for nearly a century, until the threat of bankruptcy forced their abdication, collapsing the tin market along with their infamous Palembang pits. In 1811 the British, sensing opportunity, moved to manufacture justifications for the capture of Java, then briefly under French control, and, with it, their own monopoly on the tin trade.[15] Stamford Raffles, the newly appointed lieutenant governor, was convinced that Bangka's tin was the best direction for infiltrating the reclusive Chinese market. British rule would alter the island, irrevocably binding it up in a distant world of foreign trade.[16]

Mixed with copper, tin produces bronze, bell metal, and (with zinc) gunmetal. Add lead and pewter is formed. The tin on the British Isles lay beneath rich veins of copper, and its hardness made it an ideal alloy for casting bronze.[17] But it would be Bangka's tin, in canisters filled with

5.
A joss paper seller, China, ca.1905–1910, black-and-white photo, American photographer, 20th century/private collection/ Bridgeman Images.

newly minted global commodities like tea and spice, that wrought new pathways between Europe and Asia over the next century. Bangka's tin had properties that distinguished it from the ancient (and exhausted) quarries at Cornwall. This "stream tin" was found in closer proximity to the surface. Oxidation removed those impurities common to deeper veins of cassiterite, with the river and the ocean washing away any remaining sand or silt. The result was far more supple and malleable than what remained on Britain's own shores, and it suggested a new logistics of distribution pointing to the East. The British quickly cut out the sultan and the regional administrators, opting to negotiate directly with the individual mines (*kongsi*) and their "coolie" labor.[18] Tin soon became the recognized currency of the island, and a literal currency as a constituent in Chinese coins. Although they had their own reserves, the Chinese needed ever more substantial supplies of tin to make joss paper (Figure 5), a thin foil burnt for ritual offerings. This fueled an expansive trade in a medium of exchange for which the Chinese were in constant need.[19] As Europeans became more involved in eastern competition for herbs and spices, Bangka's tin, with its unique composition, began to replace tin reserves in Europe.[20] When they returned in 1816, the Dutch state reaped the benefit of a more reliable trade with the Chinese. They too found it preferable to deal in tin because "unlike the Chinese, they did not have sufficient specialized knowledge to select high quality cargoes of [local] forest and marine products."[21] Tin had become something that was understood.

British control was brief, but it transformed the organizational structures of mining. The system was continued, with increasing Europeanization, throughout the second Dutch period. Colonial administration remained until the mid-twentieth century, ending only with the emergent Indonesian Republic's nationalization of mining operations.

Becoming Tin Men

There is little doubt that the presence of tin on Bangka came to reroute the futures of its inhabitants. Although the indigenous Bangkanese had, like other members of the Sultanate of Palembang, owed tribute, the presence of tin freed them of the burden and the established order of obligation—permitting new directions of life that had been previously impossible. The island's impressed immigrant population, on the other hand, found fortune only within the openings that had been excavated in service to the metal. Imperial mandate of the Chinese emperor prevented the legal emigration of laborers from the mainland, so transportation operated clandestinely, leaving workers in increasingly precarious positions as they sought to make their way on the island.[22] Although free Chinese sometimes came to Bangka along with Malagasy miners, most were indentured. Circular migration might occur for those who were successful. Settlement on Bangka remained for those who were not.[23]

The principal organizing structure for those who remained after British rule was the *kongsi*. These structures were initially characterized, as Mary Somers Heidhues writes, by "sanctions of mutual trust and indebtedness." For miners, the kongsi promised paid dividends, distributed individual debt, and offered a regular supply of food and equipment. They were administrative, social, and geographic compounds, built around open pits, with yards, quarters, storehouses, and places of prayer.[24] Despite their clear economic role, they remained strangely detached from the realities of value in the tin trade. Shares depended not on how much tin was produced but on the amount of work norms met.[25] Although they had initially helped to bind together otherwise fragile futures, by 1903 these organizations had regressed to little more than a systematic method of exploitation. Coolies were unable to leave without permission, and wage laborers were hesitant to become shareholders and increase their debt. The initial flexibility that had allowed locals to prosper in the success of an operation became the very mechanism through which outsiders were able to become financial speculators. As colonial mine heads subsumed collective responsibilities and common rights, these so-called little kings left the kongsi to govern operations from remote outposts.

Compared to the oppression of the late kongsi, the kind of industrial regulation brought by national control after World War II seemed, to some, a welcome reprieve. The formation of PT Timah was the first serious effort at rationalization—and, indeed, any kind of administrative reform—in nearly a century, a modern movement toward a conventionalization of previously inkonvensional practices of production. The company was certainly given ample legislative tools to attempt it. In theory, all land is worked by PT Timah, and the future of all of Bangka's tin belongs to it. PT Timah ended the reign of the little kings (Figure 6) and moved to limit the local autonomy that had produced them. The language of modernization that accompanied the

6.
"Selamat tinggal raja kecil" (Saying goodbye to the little kings) is the way to *efisiensi* (efficiency). Cover of *Stannia*, PT Timah industry magazine, 1991.

company's control heralded the coming of the *manusia timah* (literally, "tin man"). This rationalization was the new direction for mining on Bangka: a professional employee, adaptable, competitive in the global market, and loyal to the company. But, it seems, these employees must give up their hearts—their output rarely exceeds that of inkonvensional mines.[26]

Perhaps as a consequence, PT Timah remains unsure of the place for inkonvension alongside its newly minted men. The flexibility afforded the underlying materiality of tin remains enshrined in policy, law, and, above all, custom. The company has been forced to retain *kontrak* institutions for smaller mines and still issues permits to groups who continue the legacy of the kongsi system. *Koperasi Unit Desa* (KUDs, local village cooperatives) remain, as do the ad hoc structures formed through what has come to be known as "people's mining." So much compromise ensures that inkonvensional miners and tin men often work the same sites, side by side. Century-old patterns of obligation, responsibility, movement, and mining continue to persist.[27] Implicit acceptance of this mining practice has been criticized by industry observers. PT Timah's acquiescence to inkonvension, they suggest, "is like raising a tiger cub." "Once it has grown-up, it may eat you."[28]

Precarious Pathways

Bangka's landscape alternates between idyllic sands and stretches of crumbling craters where miners have plumbed nearly every inch of surface. These circular pits (30 to 50 feet deep, with enough room for up to five workers) seem to perforate the island. Little has been spared—the airport, hospital, and governor's mansion all stand within sight of modern-day Palembang pits.

The destiny of Bangka's tin, as for nearly half of all tin, is to carry the flow of electrons between the components of circuits and chipsets. Composed of more than 95% tin, with small quantities of silver and copper, solder is the connective thread of electronic communication. The number of solder points in electronic devices varies by their complexity. It can easily be upward of 5000. This number translates to a typical tin weight in smartphones of approximately 2 g per device and nearly 4 g in electronics like laptop computers. Even this small amount is already more than a quarter of the tin used in far larger productions like automobiles.[29] Indeed, when the telegraph and telephone companies first strung the wires and connected the switchboards of their communication networks at the turn of the twentieth century, they had already come to rely on the substantial material network surrounding this small Indonesian island. "Tin," Western Electric advertised in 1927, "is used as solder in your telephone and throughout telephone systems wherever wires are to be spliced." "Banka," is the "small island" that "now provides most of the tin."[30] This statement only hinted at the importance that solder would come to have over the following century, the miles of cable and the mountains of circuits for which Bangka's tin was the enabling constituent. As solder smoothed the points of connection between disparate materials, it held together not only resistors, transistors, and integrated circuits but the growth of information society itself—even as it relied on the otherwise sharp geographies and rough lives being carved out of the distant islands of the Java Sea.

The relationship between electronics production and Bangka's tin is a reliance on inkonvension in an industry whose productive practice has long belied Henry's Ford's maxim that in mass manufacturing "there are no fitters."[31] Despite images of clean-room production, high technology, and perfectly formed devices, the pathways of electronics production have been built from *within* the tenets of inkonvension, not without it. As the impacts of global manufacture direct consumer disquiet to the precarious pathways of products, avenues of this production become tinged with concerns of social abuse and environmental degradation. Yet Indonesian tin has become more, not less, desirable. In the wake of the legislative remedies that have been enacted, companies reluctant to source from Africa are redirected toward the "expedient route" to a "conflict-free guarantee." Two of Asia's largest solder suppliers, Shenmao Technology and Chernan Metal Industrial, reported buying 100% of their tin from Indonesia, selling to manufacturers like Foxconn and electronics companies like Apple, Sony, and Samsung.[32]

This demand remains even as the search for tin sand on Bangka has become difficult, its movements staggered and increasingly uncertain. Prospects have moved to the island's outer waters. Faced with a future absent its most important connection, Bangka's inhabitants have begun to draft new programs for "culture mining," as though the extractive lens is the only remaining vision for their island and its future economic security, the downward depths the only direction still imaginable.[33] Bangka's landscape has become suspect. Connection has come at a price. The scale of the environmental damage reveals a "lunar landscape of craters and hundreds of highly acidic, turquoise lakes created by centuries of largely unregulated tin mining."[34]

As scholars of globalization and production mine new sources of understanding for the uneasy relationship between man and mineral, they must continue to be cognizant that the conditions of production are neither newly emergent nor unchanging certainties. In the Java Sea, inkonvensional pathways have been painstakingly soldered into supply chains—complex assemblies of actors and groups, humans and nonhumans, each with conventions of their own.

Acknowledgments

I acknowledge the support provided by Haidy Geismar, Joshua Bell, Joel Kuipers, Carlin Wing, and Kouross Esmaeli in formulating this case study. Archival research for this work was conducted at the archives at the British Library, the National Museum of American History, and the National Museum of Natural History.

Notes

1. François Valentyn, *Oud en Nieuw Oost-Indien* (Dordrecht: Joh. van Braam, 1724).

2. The date given for the discovery of tin on Bangka is 1710. See Mary Somers Heidhues, *Bangka Tin and Mentok Pepper* (Singapore: Institute for South Asian Studies, 1992), 22.

3. Cam Simpson, "The Deadly Tin inside Your Smartphone," *Bloomberg Businessweek*, 23 August, 2012, 48–57.

4. In the *New York Times*, for example, Bangka appeared in only a few particular contexts: concerning the colonial activities of the Dutch ("Dutch Establish New Island Units," 31 July 1947), with regard to nationalization ("Indonesia Begins Nationalizing Tin," 2 March 1953), or in economic analyses concerned about the trade regulations of global tin output and the impact of tin cartels ("Indonesia Sets Survey to Spur Offshore Tin Mining," 22 February 1964, and "A Time of Crisis for Cartels," 3 December 1985). This latter frame was particularly prevalent. Even articles that discussed environmental and social impacts of tin production were initially cast in a primarily economic frame ("Indonesian Tin Industry Hits a Slump," 21 October 2008), although they are now universally associated with the manufacture of mobile devices and other consumer electronics ("Billions of Cellphones Polluting the World," 29 April 2013).

5. Arjun Appadurai, "Disjuncture and Difference in the Global Cultural Economy," *Public Culture* 2, no. 2 (Spring 1990): 6–7.

6. PT is the abbreviation for *perseroan terbatas*, the equivalent of a limited liability company, and Tbk, for *terbuka*, denotes all "public" Indonesian companies. Budy P. Resosudarmo, Ida Aju Pradnja Resosudarmo, Wijayono Sarosa, and Nina L. Subiman, "Socioeconomic Conflicts in Indonesia's Mining Industry," in *Exploiting Natural Resources: Growth, Instability, and Conflict in the Middle East and Asia*, ed. Richard Cronin and Amit Pandya (Washington, D.C.: Henry L. Stimson Center, 2009), 39.

7. Resosudarmo et al., "Socioeconomic Conflicts," 40.

8. This nomenclature takes on a particular importance when, as Kuntala Lahiri-Dutt argues, it directs how international organizations and policy makers address this kind of mining practice. Kuntala Lahiri-Dutt, "Informality in Mineral Resource Management in Asia: Raising Questions Relating to Community Economies and Sustainable Development," *Natural Resources Forum* 28, no. 2 (2004): 124.

9. In a 2006 raid the National Police forced the closure of three smelting enterprises (CV Dona Kembara Jaya, CV DS Jaya Abadi, and PT Bangka Putra Karya), seizing 84 units of illegal tin in Bangka and capturing a ship carrying 93 containers of tin bars. Illegal mining units are referred to as *Tambang Inkonvensional* (TI). Erwiza Erman, "Rethinking Legal and Illegal Economy: A Case Study of Tin Mining in Bangka Island," *Southeast Asia: History and Culture* 37 (2008): 91–111.

10. Miners can be declared illegal in a variety of ways. Although they may not hold any permit at all, they may also be accused of overreaching the mineral tenures of a legal company's "contract of work." Resosudarmo et al., "Socioeconomic Conflicts," 35–39.

11. See Erman, "Rethinking Legal and Illegal Economy," 91–111 (particularly the section "Contest of Power"), and *Stannia*, PT Timah's industry magazine, particularly *Stannia* 22, nos. 1–4 (1991), and 27, nos. 1–4 (1996).

12. As given on a 686 CE Srivijaya stone, although the word is not visible on the inscription reproduced in the lobby of the PT Timah building in Jakarta that promises "blessings on their undertakings, their clans, their families; success, health, freedom from disasters and abundance for all their lands." Heidhues, *Bangka Tin and Mentok Pepper*, 1–22.

13. See Karl Helbig, "Bangka, die 'gelbe' Insel im Malayischen Archipel," *Ostasiatische Rundschau* 21, no. 8 (1940): 158–162.

14. The ascension of the Qianlong Emperor in 1736 brought an age of prosperity and a particularly ravenous demand for imports. Lin Ken Wong, *The Malayan Tin Industry to 1914, with Special Reference to the States of Perak, Selangor, Megri Sembilan and Pahang* (Tucson: University of Arizona Press, 1965), 4, 218.

15. For an overview of the complex interplay of commercial, imperial, and indigenous forces in Indonesia in the eighteenth and nineteenth centuries, see M. C. Ricklefs, *A History of Modern Indonesia Since c. 1200*, 3rd ed. (Stanford, CA: Stanford University Press, 2001).

16. Lord Minto to Stamford Raffles, 15 December 1812. See the records of the occupation of Bangka, India Office Records, IOR G/21/60, British Library.

17. An ancient source of the metal. Pliny the Elder wrote of "tin islands," called by the Greeks "Cassiterides, in consequence of their abounding in tin." He likely refers to the Scilly Isles near Cornwall, but the geography is disputed. Greek and Roman geographers tended to borrow from the stories told by Phoenician merchants (for whom tin was a trade secret). The Cassiterides

are also referenced by Herodotus and Diodorus. See Pliny the Elder, *The Natural History*, trans. John Bostock and H. T. Riley (London: Taylor and Francis, 1855), book 4, chap. 36. Strabo particularly muddles these locations, writing that the Cassiterides are "ten in number, lying near each other in the ocean, towards the North from the haven of the Artabri." In the ancient geography the Cassiterides become an imaginary mixture of the Azores and the Scilly Islands. Strabo, *Geography*, trans. H. L. Jones (Cambridge: Loeb Classical Library, 1923), vol. 2, book 3, chap. 5.

18. India Office Records, IOR G/21/60, British Library.

19. E. S. Hedges, *Tin in Social and Economic History* (London: Edward Arnold, 1964), 94. See also Stamford Raffles to Lord Minto, 1 August 1812, India Office Records, British Library.

20. Leonard Andaya, *The Kingdom of Johor* (London: Oxford University Press, 1975), 68–69.

21. Barbara Andaya and Leonard Andaya, *A History of Malaysia* (New York: Palgrave Macmillan, 1982), 93.

22. Drawn from J. A. Schuurman's reading of a 1770 account. Heidhues, *Bangka Tin and Mentok Pepper*, 8–10.

23. "They feel obligated to work for two years according to rules that bind them, and the position of dependence, which they have taken on voluntarily, is, viewed from their own standpoint, nothing other than a temporary slavery. They speak of re-engagement, for which they receive a premium of f35, as 'selling oneself' [*mai shen*], without finding anything unnatural or demeaning in this expression which sounds so unpleasant to our ears." De Kat Angelino, 1919, quoted in Heidhues, *Bangka Tin and Mentok Pepper*, 124, see also 110–124.

24. In this respect, the kongsi emerges as a more durable form of the cooperative mining operations found in sites like Cornwall. Heidhues, *Bangka Tin and Mentok Pepper*, 39. See also Mary Somers Heidhues, "Company Island: A Note on the History of Belitung," *Indonesia* 51 (1991): 7.

25. These work norms were units of soil dug and carried, regardless of the presence of tin. "The shares [*hun*] in a mine are usually divided equally among the miners; however there are some private Chinese, whether traders or other *kampung* [town]-dwellers, who take or purchase a couple of shares in a mine and have it worked by coolies who are paid a wage." Fraenkel writing in 1843, quoted in Heidhues, *Bangka Tin and Mentok Pepper*, 40.

26. Agriculture remains important, but since PT Timah owns the rights for all future mineral extraction on the island (outside of towns and villages), land can be repossessed at any time, and ventures carry significant risks. Heidhues, *Bangka Tin and Mentok Pepper*, 214. See also Heidhues, *Bangka Tin and Mentok Pepper*, 72–76; Kemas Kurniawan, "The Post-crisis Indonesian Tin Town: With Reference to Mentok, Bangka," *International Journal of Environmental, Cultural, Economic, and Social Sustainability* 5 (2005): 90.

27. Until recently, there were no legal codes and regulations in place for small-scale miners other than the mining law of 1967, and much of their practice was governed by custom and industry fiat. Article 11 of that law (titled "Peoples Mining") states that "the objective of Peoples mining is to give the local population opportunity to exploit minerals in their efforts to participate in the development of the state in the field of mining under the guidance of Government." Clive Aspinall, "Small-Scale Mining in Indonesia," *Mining, Minerals and Sustainable Development* 79 (September 2001): 11.

28. Hartono, P. D. Prabandari, and Asnadi, "Biting the Hands That Feed," *Tempo Magazine*, September 2001, 62–63.

29. One inkonvensional miner, Mistar, helps contextualize this in terms of individual output: "I work with a friend. We can collect 7 kg of tin a day—that's worth about 490,000 Indonesian rupiahs (about $50). Friends of the Earth, *Mining for Smartphones: The True Cost of Tin* (London: Friends of the Earth, 2012), 17.

30. Western Electric, *From the Far Corners of the Earth* (New York: McGraw Hill, 1927), 34–35.

31. David Hounshell, *From the American System to Mass Production, 1800–1932: The Development of Manufacturing Technology in the United States* (Baltimore: Johns Hopkins University Press, 1985), 107.

32. Simpson, "Deadly Tin," 52–53.

33. "Culture mining" is a phrase appearing in local initiatives and workshops (such as the mAAN Mentok Tin City Design Workshop in 2011), which could be read to advocate drawing out future economic value from the island's history and culture.

34. Ed Davies and Fitri Wulandari, "Indonesia's Tin Islands: Blessed or Cursed?" Reuters, 21 October 2008, http://reut.rs/h5GjeX (accessed 24 January 2010).

Bibliography

Appadurai, Arjun. "Disjuncture and Difference in the Global Cultural Economy." *Public Culture* 2 (1990): 1–23.

———. *Modernity at Large: Cultural Dimensions of Globalization*. Minneapolis: University of Minnesota Press, 1996.

Andaya, Barbara, and Leonard Andaya. *A History of Malaysia*. New York: Palgrave Macmillan, 1982.

Andaya, Leonard. *The Kingdom of Johor*. London: Oxford University Press, 1975.

Aspinall, Clive. "Small-Scale Mining in Indonesia." *Mining, Minerals and Sustainable Development* 79 (September 2001): 1–30.

Davies, Ed, and Fitri Wulandari. "Indonesia's Tin Islands: Blessed or Cursed?" Reuters, 21 October 2008. http://reut.rs/h5GjeX (accessed 24 January 2010).

Erman, Erwiza. "Rethinking Legal and Illegal Economy: A Case Study of Tin Mining in Bangka Island." *Southeast Asia: History and Culture* 37 (2008): 91–111.

Friends of the Earth. *Mining for Smartphones: The True Cost of Tin*. London: Friends of the Earth, 2012.

Hartono, P. D. Prabandari, and Asnadi. "Biting the Hands That Feed." *Tempo Magazine*, September 2001, 62–63.

Hedges, E. S. *Tin in Social and Economic History*. London: Edward Arnold, 1964.

Heidhues, Mary Somers. *Bangka Tin and Mentok Pepper*. Singapore: Institute for South Asian Studies, 1992.

———. "Company Island: A Note on the History of Belitung." *Indonesia* 51 (1991): 1–20.

Helbig, Karl. "Bangka, die 'gelbe' Insel im Malayischen Archipel." *Ostasiatische Rundschau* 21, no. 8 (1940): 158–162.

Hounshell, David. *From the American System to Mass Production, 1800–1932: The Development of Manufacturing Technology in the United States*. Baltimore: Johns Hopkins University Press, 1985.

India Office Records. British Library.

Jackson, James. "Mining in 18th Century Bangka: The Pre-European Exploitation of a 'Tin Island.'" *Pacific Viewpoint* 10, no. 2 (1969): 28–54.

Kurniawan, Kemas. "The Post-crisis Indonesian Tin Town: With Reference to Mentok, Bangka." *International Journal of Environmental, Cultural, Economic, and Social Sustainability* 5 (2005): 86–106.

Lahiri-Dutt, Kuntala. "Informality in Mineral Resource Management in Asia: Raising Questions Relating to Community Economies and Sustainable Development." *Natural Resources Forum* 28, no. 2 (2004): 123–132.

Pliny the Elder. *The Natural History*. Translated by John Bostock and H. T. Riley. London: Taylor and Francis, 1855.

Resosudarmo, Budy P., Ida Aju Pradnja Resosudarmo, Wijayono Sarosa, and Nina L. Subiman. "Socioeconomic Conflicts in Indonesia's Mining Industry." In *Exploiting Natural Resources: Growth, Instability, and Conflict in the Middle East and Asia*, ed. Richard Cronin and Amit Pandya, pp. 33–46. Washington, D.C.: Henry L. Stimson Center, 2009.

Simpson, Cam. "The Deadly Tin inside Your Smartphone." *Bloomberg Businessweek*, 23 August 2012, 48–57.

Stannia 22, nos. 1–4 (1991); 27, nos. 1–4 (1996).

Strabo. *Geography*. Translated by H. L. Jones. Cambridge: Loeb Classical Library, 1923.

Valentyn, François. *Oud en Nieuw Oost-Indien*. Dordrecht: Joh. van Braam, 1724.

Western Electric. *From the Far Corners of the Earth*. New York: McGraw Hill, 1927.

Wong, Lin Ken. *The Malayan Tin Industry to 1914, with Special Reference to the States of Perak, Selangor, Megri Sembilan and Pahang*. Tucson: University of Arizona Press, 1965.

Wubin, Zhuang. *Chinese Muslims in Indonesia*. Jakarta: Select Publishing, 2011.

Dimensions of Globalization: Objects and Identities

CHAPTER 6

Local Makers, Global Players
Tabla Manufacture and Design in a Global Marketplace

P. Allen Roda
Jane and Morgan Whitney
Research Fellow

Department of Musical Instruments
Metropolitan Museum of Art
New York, New York

The growing popularity of what has been commonly referred to as "world music" is both an indication and product of globalization. This popularity has led to greater demand not only for Indian music recordings and musicians overseas but also for Indian musical instruments themselves—particularly a north Indian set of drums called tabla. Tabla are harmonically complex and tonally rich sets of drums used in a variety of musical genres ranging from the devotional music of Hindus, Sikhs, and Muslims to film, folk, and fusion. They are perhaps most famous for their role in Hindustani classical music as accompaniment and also as a solo instrument. The drums' unique design has developed through close interactions between performers and instrument makers since the drums began to be widely circulated approximately 200 years ago.[1] Although South Asian music has been growing in international popularity, tabla are also increasingly incorporated in musical genres unrelated to South Asia, creating a further rise in the demand for the instrument and, consequently, changes in local tabla economies.[2] As Arjun Appadurai noted, globalization results in neither the "triumphantly universal" nor the "resiliently particular," but rather the movement of people, money, and objects through various flows he refers to as "scapes."[3] As the tabla moves through these various and increasingly global flows over the course of its "social life,"[4] it changes from specialized item to commodity and back,[5] constantly forging and influencing new relationships with and between humans.[6]

Using participant observation with tabla makers in Banaras (also called Varanasi) and oral histories of the trade, I analyze various ways in which tabla makers engage with new opportunities and constraints placed upon them by international markets, including changes in maker-customer interactions, the growth of wholesale manufacture, and changes in the tabla's design and materials. I examine the global tabla industry within the framework of "detraditionalization" as part of a globalization narrative that highlights a breaking of norms in pursuit of new opportunities generated by increasingly open international markets and prevailing neoliberal models of capitalism.[7] In this case study, I trace the relationship between changing global trends in music consumption and international relations between countries (Australia and India) to the business practices of Banarasi tabla makers. Tabla are physical objects through which personal, musical, financial, and increasingly international relationships are constituted. Critical analysis of their movements through different flows or scapes yields insight into the relationship between global trends, national policies, and local practices as well as the way in which these resounding objects bring people together across national, cultural, and linguistic barriers.

Increased circulation of tabla players, tabla students, and Indian music over the course of the twentieth century has resulted in a vibrant international community of tabla enthusiasts, with players and teachers on every continent.[8] These tabla players continue to acquire their drums from India, as tabla makers have not yet emigrated.[9] Some make semiannual trips to visit their tabla maker or request orders from friends or family who will be traveling, whereas others depend upon long-distance relationships mediated by electronic communications. Some experienced players and many new students depend on retailers who act as middlemen. These retailers have created the market for "wholesale" tabla distribution that has had a powerful effect on the tabla economy of Banaras, drastically increasing the demand for the instrument. Three of the largest musical instrument retailers in Delhi (BINA, DMS, and Lahore Musical Instruments) reported between 200% and 300% growth in tabla exports in the last 10 years.[10]

Tabla Technology and the Role of Maker-Player Interaction

The set of drums consists of a larger metal drum that is hand forged from brass or copper and a smaller wooden drum that is rounded on a lathe and carved out by hand.[11] The two drum shells are covered by multiple layers of goat rawhide that are attached by woven rings of buffalo rawhide and straps. On top of the drum heads, tabla makers apply a tuning paste, which is a mixture of dough (wheat cooked with water) and a powder made of iron slag or stone (Figure 1). The tuning paste is applied in thin layers fractions of a millimeter in thickness that are then polished with stone, forcing the drum head to bend while it dries. Bending the skin causes tiny cracks to form in the paste while it dries, creating individual grains that are each attached to the drum head but separate from each other. Up close they look like pieces of a very tiny jigsaw puzzle (Figure 2). The tuning paste adds weight to the drum head, giving the instrument its distinct ringing sound,

1.
A standard tabla set: nickel-plated brass and wooden resonators with composite heads made of four layers of goat rawhide and tuning paste, called *syāhi*. Surrounding each head is a woven ring of buffalo rawhide, and the heads are attached with buffalo rawhide straps. Photo by the author.

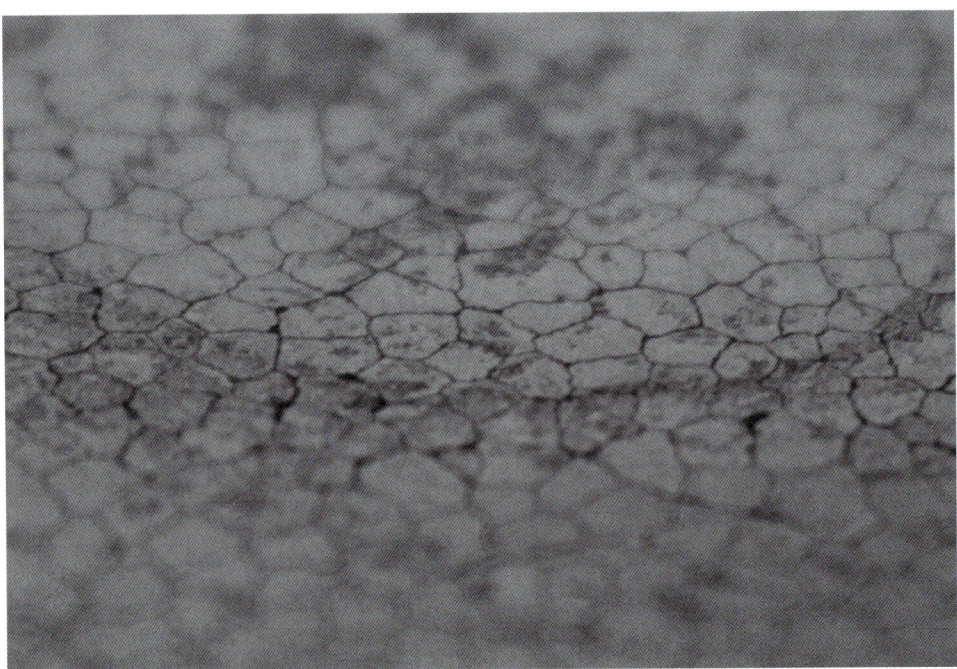

2.
Close-up of the tuning paste, called *syāhi*. Notice the tiny cracks that form from polishing the head with a stone. The tuning paste adds weight to the drum head, lowering its pitch and increasing its sustain. The cracks allow each grain to vibrate independently, preventing the tuning paste from dampening the resonance of the instrument. The way in which the *syāhi* is applied is central to each maker's personal style. Photo by the author.

and the cracks enable each grain to vibrate independently and therefore limit the paste's dampening effect on the main skin.

All tabla are not created equal, and variations in materials and craftsmanship can have important impacts on the sounds they are capable of making. Banarasi tabla are famous for having thicker skins and thicker tuning paste, which helps to give them their signature sound—usually described as being "full" and "resonant," in contrast to tabla made on thinner hides, whose sound is often called "bright."[12] Although all of the stages of tabla manufacture are important, making subtle adjustments to the shape of the tuning paste (by adding or removing strategically placed layers) can change the sound of an instrument in significant ways. During this stage, most serious tabla players choose to be present with the tabla maker, listening to these changes and helping to decide when it is finished. It is not a guarantee that further adjustment will improve the sound, so it is important for players and makers to be able to agree that it is "done." Also, tabla may sound differently in one player's hands than in another's, as variations in training and experience become important when shopping for the right sound. Consequently, players often contribute actively to the fine-tuning of their instruments. It is usually not a quick process and can take anywhere from 30 minutes to several hours, involving at least one cup of *chai*, some *paan*, and small talk.[13]

Furthermore, every player and maker I interviewed agreed that the drum heads should be built directly on the shells so that the maker can adjust the head accordingly. Since the drum heads get damaged or stretched over time, players must visit their makers frequently for repairs and replacements.[14] Patron-client relationships vary, with some players being extremely loyal to a particular tabla maker while others think of them as inconsistent, fraudulent, and deceitful. Whether friendly or strained, the relationship between maker and player involves an intimacy that comes from close listening to the instrument, intense discussions of sound, and negotiations of price and quality.

Wholesale Karna

The growth of tabla exports to other countries has given tabla makers access to new clientele that disrupts the player-maker relationships described previously. Many Banarasi tabla makers work almost exclusively for Delhi-based retailers instead of local customers. In Banaras, this practice is referred to as "wholesaling."[15] Wholesaling has played the primary role in facilitating the drastic increase in tabla makers employed in Banaras. In the last 25 years, Banarasi tabla makers have grown from about seven or eight makers in a few shops to 75 or more makers today, with not enough sons born to meet the growing demand for the instruments.[16] Two cousins (Bashir-ud-din and Shams-ud-din) who worked together in the same shop 25 years ago now have a total of 12 sons who each have their own tabla shops staffed primarily with family members but also with hired help. Two grandchildren of those same cousins are now going to college to get business degrees and are shifting away from artisanal craftsmanship toward import/export management because their family's wholesale business has grown so much that they do not need to make tabla themselves.[17]

Whether they are oversimplifying the past or lamenting the current state of affairs, two tabla makers with whom I worked closely suggested that in previous generations, success depended on the quality of the drums one produced rather than the quantity.[18] This transition can be seen as part of a larger pattern of detraditionalization as outlined by Ulrich Beck, Scott Lash, and Anthony Giddens in which new opportunities lead individuals away from traditional modes of production.[19] When I asked one wholesaler, Niyaz Ahmed, about that choice, he told me that he would rather make drums than spend the afternoon haggling with a customer. Although the profit margin per drum is less, he argued that he can make more drums per day because he is free from having to deal with customers.[20] On separate occasions, four different Banarasi tabla makers accused wholesalers of "making jalebis," which is applying tuning paste in concentric rings to make the drum head look attractive without putting any effort into balancing the drum's harmonics, opening up its voice, or otherwise trying to improve its sound quality.[21]

In this sense, wholesaling is a disruption of the traditional collaborative effort of players and makers that typifies production patterns for local customers. Fred Myers refers to the notion of "regimes of value" when he writes of the difficulty of translating the religious truth value of Pintupi paintings to their aesthetic value as works of modern art.[22] Similarly, when players and makers work together to produce a unique sound, the commodity they exchange is the sonic capacity of the instrument. That is, the player shops for the drum's sound, not for the drum. When drum heads are removed from their shells and sent silently to Delhi to be reattached, repackaged, and shipped overseas, the drum heads themselves are the commodity, not their sound, because retailers do not have the opportunity to be so closely invested in the fine-tuning process. Separating the head from the shell also makes important aspects of fine-tuning impossible.

The sheer volume of tabla heads retailers receive would also make careful scrutiny of each one's sound very difficult, if not impossible (hence the increased importance of visible indicators like the concentric rings). Two different retailers complained that after contracting a tabla maker, they would often find that the slowly deteriorating quality of shipments would result in more and more complaints from customers, until ultimately they felt compelled to work with someone else.[23] At the retail shop, instruments are assembled, packaged, and then sold as part of a line of drums that represents a particular company's "standard models," where any reference to the instrument's unique personality or the style of the maker is further erased. In many ways, this model of mass production and branding is an accurate reflection of movement along George Ritzer's spectrum from "something" to "nothing" (i.e., from inalienable, unique objects to ubiquitous, standardized ones).[24] This shift also reflects Igor Kopytoff's description of "commoditization as process."[25] Mass production, or wholesaling, marks a change in tabla technology as the evaluation of the instrument is based less on a sonic "regime of value" and more upon the reputation of the maker and visual aesthetics.[26]

Artisanal Craftsmen and International Markets

Foreign visitors and students passing through town offer tabla makers a different kind of opportunity than local customers. Since they know that these customers are unlikely to return to the shop, there is little incentive for tabla makers to spend exorbitant amounts of time tuning their drums. Furthermore, the students themselves often do not have the skills or experience to evaluate drums for sound quality. Those who have ever tried to learn to play tabla will know that it takes weeks if not months of training before new students can even make some of the basic tabla sounds, and even then, they would not be consistent enough in their hands to know whether variations in sound were related to the drum or to their playing. As master craftsman Mohammed Anwar told me, "If the customer is happy, I am happy," implying that the customer was the ultimate arbiter as to whether or not the instrument was "finished" and that there was no need to work toward improving the sound of an instrument if the customer was satisfied.[27] This rubric is the same for advanced players as it is for novices; however, with novices it seems to take on a different meaning because it is presumed that they won't know what to listen for or how to do it.[28] New markets mean new opportunities, and many tabla makers learn to adjust their business practices in order to benefit from these new markets as much as possible.

Of course, not all foreign customers are tourists just passing through. Many students in Banaras return year after year to further their musical training and often buy multiple drums with each visit. Some stay for weeks, others months.[29] I met two such students who are now teachers in their home countries and import tabla for themselves and their students with some regularity. In situations like these, the players prefer to check their drums during the fine-tuning process just like serious local performers do, but often, they do not have the opportunity.[30]

Although relatively few artisanal tabla makers regularly ship drums overseas, doing so has forced them to adjust their business practices to include various electronic media for communication and to account for bureaucracies surrounding international shipments, money transfers, and customs. One such tabla maker is Mohammed Anwar, whose attempts to ship drums to a customer, Shen Flindell, in Australia have also led to changes in the instrument's design and, ultimately, the creation of an "Australia model tabla" with synthetic component parts that can pass through the Australian quarantine without irradiation. Although globalization has affected tabla technology obliquely through the growth of the wholesale market and changing dynamics between tabla makers and visiting foreign musicians, the intervention of the Australian Quarantine and Inspection Service impacts tabla technology directly through forced incorporation of new materials and structural changes to the instrument.

The quarantine has very strict, yet vague, rules about the types of animal products that can be allowed into the country and whether or not these products must be subjected to gamma irradiation treatment (at 1,600 times a lethal dose for humans) before they can enter.[31] Irradiation ruins tabla heads. They can no longer hold tension, the straps often disintegrate, and the heads may simply fall apart or crumble to dust. Since 1998, Flindell has tried a variety of methods to get around irradiation. When he accompanied his luggage, he could occasionally talk his way into getting them through customs, although it was always a risk, and he has been denied entry on multiple occasions. He began sending the metal and wooden shells separately and hiding the leather goods in bed sheets and other textiles so that they might pass unnoticed. One Australian tabla player on an online forum deliberately ordered his drums from Canada because he thought that shipments from Canada would be less scrutinized (even though the drums were still made in India).[32] Similarly, Flindell would sometimes ship his drum heads from Japan.[33]

On the basis of his experiences of some tabla passing inspection and others not, through the years, Flindell surmised that tabla were more likely to pass through customs if the heads were white, clean looking, and devoid of any hairs. For a brief period, Anwar tried painting the top layer of the drum head with white nail polish to see if it would help. It didn't. Since the Australian quarantine's website did not specify what criteria would allow untanned animal skins to pass, Flindell researched a report that outlined the types of diseases that concerned the quarantine and discovered that these diseases would only travel on thick pieces of rawhide rather than thin ones. Subsequently, the main goat hide heads were safe; it was the woven ring of buffalo hide and the straps that were the primary problem. It took several incarnations of Australia model drum heads made with various types of synthetic materials before Anwar finally found a combination that would both pass through the quarantine without irradiation and produce satisfactory sounds (Figure 3).

Anwar used plastic straps from shipping large appliances and imitation-wood window frame molding, drastically altering the tabla's design (and therefore its technology). These materials do not allow him to put as much tuning paste on the drums as he would use on a traditional tabla and thus do not perfectly replicate his signature sound. He also developed new manufacturing methods in order to incorporate these materials that deviate from traditional practices in significant ways. The Australia model tabla is a unique invention by Mohammed Anwar driven exclusively by the increased international market for tabla and Australia's trade and quarantine relations with other countries. There are no intellectual property laws that he would be able to enforce should someone copy his design, and his primary protection is the fact that these drums do not circulate in Banaras—only in Australia. Consequently, he has asked me not to publish the details of the techniques involved in producing these instruments.

Although Flindell is relatively satisfied with the performance of Anwar's latest Australia model, he still prefers the look, feel, and sound of traditional Banarasi tabla.[34] Thin wisps of plastic stick out from the top ring of the drum that is the primary intervention of the Australia model. Some of Anwar's customers have complained that the plastic sheds and can cause splinters when they play for extended periods of time. He advises them to be careful not to touch the outer ring while practicing, which consequently affects the way they approach the instrument, as most beginners and even many advanced players are likely to rest their palms on this part of the drum during performance.[35]

Flindell finally found a long-term solution that will allow him to import traditional Banarasi tabla on a regular basis without subjecting them to irradiation. From his research, he convinced the quarantine that the drum heads are safe for import by documenting that the hides have been treated with calcium hydroxide (lime water) at a pH level of 12.5 or higher. After several failed trips to government offices in Kanpur and Lucknow to meet various export authorities in Uttar Pradesh, he was finally put in contact with someone on the veterinary board in Banaras who was

3.
Mohammed Anwar's "Australia model" tabla. Notice the straps and the woven ring surrounding the head are made of synthetic materials. These materials cannot support the same tension as buffalo rawhide, thus changing the techniques of both tabla makers and tabla players as they adjust to using lighter-weight tuning paste. Also, little wisps of plastic sometimes shed from the woven ring and can be bothersome to performers' hands. Photo by the author.

able to certify the safety of the products. Since October 2012, Flindell has regularly received parcels with the new certification, so he no longer orders partially plastic tabla.[36] The Australia model tabla is now a rare entity known only to Australian tabla players of the last few years.

Conclusion

Tabla making has always followed the needs of tabla players and the market. Regional variations in design reflect similar variations in performance practice.[37] As the world of tabla performance grows beyond India's borders, makers continue to innovate and adjust to the specific needs of their customers, taking advantage of new business opportunities whenever possible. It is difficult to say what these trends portend for the future. The demand for increasingly cheaper drum sets shipped all over the world is growing, while the number of players who wish to participate in the construction of their instruments and maintain a personal relationship with their makers stays steady. Tabla makers like Mohammed Anwar and Ekhlak Ahmed complain that their customers want them to spend extra time fine-tuning their instruments but then pay a wholesale rate.[38] If they are unable to raise rates sufficiently to compensate for the extra effort it takes to fine-tune their instruments, they may be forced to work toward producing a higher quantity of lower-quality drums in order to earn a living wage.

Increasing direct engagement between tabla makers and foreign customers may help them maintain a sufficient client base at high enough rates to stay in business; however, doing so complicates business practices drastically. Anwar, for example, depends heavily upon his seventeen-year-old son, Salman, who has learned some English in school and (through interactions with the author and Flindell) has a working knowledge of computers and the internet. Although communication technologies now enable the long-term personal relationship between players and makers to extend beyond local customers, there are still limitations. Computer-aided telecommunication services, like Skype or Google Hangouts, cannot replace the multisensory experience of listening to a new drum played by one's own hands.

Notes

1. Kippen points out that the extensive iconographic references to other types of drums prior to 1745 and the absolute lack of any iconography of tabla prior to that date suggest that if tabla existed prior to 1745, they certainly were not popular in the courts at that time. James Kippen, "The History of Tabla," in *Hindustani Music: Thirteenth to Twentieth Centuries*, ed. Joep Bor (Delhi: Manohar, 2010), 460.

2. Fusion music was once limited to jazz ensembles like Zakir Hussain and John McGloughlin's group Shakti and Karsh Kale. However, artists such as Harry Manx and Hindugrass are increasingly fusing Indian music styles with American folk, country, and bluegrass traditions. Talvin Singh, Buddha Bar, and Thievery Corporation are excellent examples of performers who incorporate tabla into electronic dance music genres.

3. Arjun Appadurai, "Disjuncture and Difference in the Global Cultural Economy," *Public Culture* 2, no. 2 (March 1990): 24.

4. Here I refer not only to Arjun Appadurai's introduction to the edited volume *Social Life of Things* but also to Eliot Bates's recent article "The Social Life of Musical Instruments." Bates incorporates more recent theories of material culture (such as actor-network theory and Jane Bennett's notion of "thing power") into a concept of "social life" that places greater emphasis on what the objects do to the humans they engage with rather than simply drawing attention to the myriad relationships humans have with objects. For further insight into the connection between musical instrument studies and theories of material culture, see Eliot Bates, "The Social Life of Musical Instruments," *Ethnomusicology* 56, no. 3 (2012): 363–395; for an excellent discussion of the "vital materiality" of objects and their ability to influence outcomes, see Jane Bennett, *Vibrant Matter: A Political Ecology of Things* (Durham, NC: Duke University Press, 2009).

5. Igor Kopytoff, "Cultural Biography of Things: Commoditization as Process," in *The Social Life of Things: Commodities in a Cultural Perspective*, ed. Arjun Appadurai (Cambridge: Cambridge University Press, 1986), 64–94.

6. Following the example of Bruno Latour in *We Have Never Been Modern* and *Reassembling the Social* and Jane Bennett in *Vibrant Matter*, I shift my focus on the study of musical instruments away from what the objects *are* and focus rather on what they *do* and subsequently treat them as nonhuman actors rather than passive recipients of human agency. Bruno Latour, *We Have Never Been Modern*, trans. Catherine Porter (Cambridge, MA: Harvard University Press, 1993); *Reassembling the Social: Introduction to Actor Network Theory* (New York: Oxford University Press, 2005).

7. Ulrich Beck, Scott Lash, and Anthony Giddens, *Reflexive Modernization: Politics, Tradition and Aesthetics in the Modern Social Order* (Stanford, CA: Stanford University Press, 1994); Chamsy El-Ojeili and Patrick Hayden, *Critical Theories of Globalization: An Introduction* (New York: Palgrave Macmillan, 2006).

8. World of Tabla lists 222 tabla teachers in the United States alone; http://world-of-tabla.com/tabla_teacher_usa.php (accessed 17 October 2012).

9. There are many explanations for why tabla makers themselves have not emigrated overseas, including but not limited to access to travel documents and the financial capacity to travel internationally, subsequent difficulty acquiring the raw materials for tabla manufacture outside of India, and competing with Indian tabla makers. Although the shipping costs would go down, the other costs of business, such as rent, would make it nearly impossible for a tabla maker to survive economically in the United States or Europe.

10. Interviews conducted with DMS, BINA, and Lahore Musical Instruments (April 2011).

11. On cheaper models, steel and even aluminum can be used, although they are not manufactured in Banaras. There are still a few clay shells available, although their fragility makes them rare. Various types of wood are also available, and outside of Banaras machines are occasionally used to drill open the shells of cheaper drums.

12. In chapters 4 and 5 of my dissertation, "Resounding Objects: Musical Materialities and the Making of Banarasi Tablas" (Ph.D. diss., New York University, 2013), I detail each of the processes involved in tabla making, the pitfalls caused by inconsistent materials that tabla makers must frequently overcome, and the creativity, dexterity, and musicality of tabla makers to improvise solutions to problems that almost invariably emerge in tabla making. I also analyze the craftsmanship of different makers through recordings and spectral analysis to try and articulate quantifiable differences between instruments. Every tabla player and maker I encountered confirmed the notion that all tabla are not made the same and that the materials and craftsmanship involved in tabla making played an important role in the sounds the instrument was capable of producing.

13. These observations come from apprenticeships with Mohammed Anwar and Ekhlak Ahmed from 2010 to 2011, during which time I learned to make my own instruments and also closely monitored interactions between customer and client while working at the tabla shop. For an extensive discussion of fine-tuning, see P. Allen Roda "Tabla Tuning on the Workshop Stage: Toward a Materialist Musical Ethnography," *Ethnomusicology Forum* 23 (2014): 360–382, doi:10.1080/17411912.2014.919871.

14. Of the eighteen Banarasi tabla players I interviewed during my fieldwork, all of them owned multiple sets of drums, and all but one of them had at least one drum currently in a shop somewhere for repair, making return trips seem more like visiting a doctor or a mechanic than a retail shop.

15. In local parlance, *wholesale karna* has become a verb that refers to working toward mass production for large clients, rather than small-scale production for individuals. Consequently, *wholesale karnewala* is the expression used to describe someone who works in this way. The experience of Banarasi wholesalers can be compared to those in other cities; however, there are no cities in India that have as many tabla makers per capita as Banaras. Consequently, the Banarasi situation is also unique in many ways.

16. The oral history I have gathered surrounding tabla making in Banaras comes from multiple formal and informal interviews conducted between 2010 and 2011 with six tabla makers (Anwar, Ekhlak, Munna, Shamshuddin, Waseem, and Mumtaz) representing Banaras's well-established tabla making families. I cannot confirm it as fact, but I would suggest that with very few potential exceptions, nearly all of Banarasi tabla makers today are related to members of these families through marriage, kinship, or mentorship. I have cross-referenced their statements about previous generations to put together a genealogy of tabla making that demonstrates the interrelatedness of the industry in Banaras. As *khandāni kām*, or "family work," tabla making is traditionally passed from father to son, with tabla makers only reluctantly contracting laborers from outside the family. Although women do contribute substantially to tabla making, they do so from their homes, as many maintain *pardah* and avoid encounters with men to whom they are unrelated. Consequently, there are no female tabla shopkeepers in Banaras. For a genealogy of Banarasi tabla makers, see chapter 6 of my dissertation, "Resounding Objects."

17. Iqbal Ahmed and his two sons, Arif and Arshad, interview with the author, 13 March 2011.

18. Mohammed Anwar and his brother Mumtaz on separate occasions expressed frustration that the practice of wholesale was keeping both the quality and the price of tabla down and making it difficult to raise their rates. They suggested that this practice is relatively new and that in their father's generation no one could make a living by shipping large orders to Delhi. Interviews with the author conducted in March 2011.

19. Beck et al., *Reflexive Modernization*, 79.

20. Niyaz Ahmed, personal communication, February 2011.

21. Anwar, Ekhlak, Mumtaz, and Imtiyaz all specifically referred to these concentric rings derogatorily as *jalebis*, which is the name of a ubiquitous spiral shaped sweet in North India. Imtiyaz Khan said, "Here we don't make *jalebis*, we make the *lov* (the center of the drum head) and the *chati* (the edge of the drum head) balanced. Take *jalebis* if you want beauty. If you want voice, don't take *jalebis*" (interview with the author, 4 March 2011). Whether or not a drum head has concentric rings is not a reliable indicator of its sound quality, as there are well-known tabla makers outside of Banaras who use concentric rings while also balancing and opening the drum's voice. Banarasi tabla makers, however, suggest that this is also a technique that tabla makers use to sell drums that have not been tuned.

22. Fred Myers, "Wizards of Oz: Nation, State, and Production of Aboriginal Art," in *Empire of Things: Regimes of Value and Material Culture*, ed. Fred Myers (Santa Fe, NM: School of American Research Press, 2001), 165–204.

23. These two retailers will remain anonymous because I do not want to publicize that they periodically receive inferior quality products and potentially harm their business.

24 George Ritzer, *The Globalization of Nothing 2* (Thousand Oaks, CA: Pine Forge Press, 2007), 40.

25. Kopytoff, "Cultural Biography of Things," 64.

26. Myers, "Wizards of Oz," 165–204.

27. Mohammed Anwar, interview with the author, March 2011.

28. One tabla player recounted tales of watching his tabla maker deliberately hide the defects of a particular tabla in order to sell it more quickly to less knowledgeable customers. Interview with the author, 13 March 2011.

29. Having spoken with ten international tabla students, both studying at Banaras Hindu University and undertaking private tutelage, I learned that there is actually a rather vibrant expat community of primarily music students who return each year for the "season," which is roughly from September to April. Many of them know each other, practice together, and often live in the same guesthouses. One sitar student told me that he can earn enough picking fruit in France each summer to pay for his accommodation and lessons for nine months in Banaras.

30. I met one foreign tabla player in the United States who usually orders several tabla at once, chooses the one that suits him the best, and then passes the others on to students. He has even asked me to help him adjust the instruments he receives to see if they can be made more suitable because he is frustrated by the unpredictability of the shipments. He will remain anonymous because I would not want his students to get the impression that they are playing inferior drums.

31. According to permits issued to Flindell and Anwar under the Australian Quarantine Act, Section 13 2AA, and Import Case Details—Public Listing by the Australian Quarantine and Inspection Service, gamma irradiation treatment for untanned hides and skins (all species excluding crocodile skins) is at a minimum of 50 kilograys (kGy). Australian Quarantine and Inspection Service, "Import Case Details—Public Listing," http://www.theQuarantine.gov.au/icon32/asp/ex_casecontent.asp?intNodeId=8198614&intCommodityId=13334&Types=none&WhichQuery=Go+to+full+text&intSearch=1&LogSessionID=0 (accessed 3 April 2012, site discontinued). Further information on the Australian Quarantine may be found here: https://www.comlaw.gov.au/Details/F2012C00627/Html/Volume_1#_Toc334790608). Merck Pharmaceuticals suggests that human exposure to 30 Gy gamma radiation results in 100% chance of death within 48 hours. Jerold T. Bushberg, "Radiation Exposure and Contamination," Merck Manuals, last modified June 2009, http://www.merckmanuals.com/professional/injuries_poisoning/radiation_exposure_and_contamination/radiation_exposure_and_contamination.html?qt=&sc=&alt= (accessed 3 April 2012).

32. Username reemixx on the popular tabla forum chandrakantha.com, posted 11 August 2009, http://forums.chandrakantha.com/viewtopic.php?f=3&t=6306&hilit=Reemix#p37296 (accessed 14 October 2012). His drums ultimately traveled three-fourths of the earth's circumference before reaching him in Perth.

33. Multiple interviews by the author with Shen Flindell (February and March 2011) and ongoing electronic communication via email, Skype, and Facebook.

34. Banarasi tabla are famous for the use of thicker, heavier tuning paste to get their characteristically full, resonant tone. The synthetic ring around the drum head is not as strong as the traditional buffalo hide ring, and consequently, Anwar must apply less tuning paste and adjust his tuning methods to compensate. As a professional performer and tabla teacher with over 20 years of experience, Flindell is very sensitive to these changes.

35. Flindell, personal communication, June 2012.

36. Flindell, personal communication, June 2014.

37. Regional designs also influence performance practice as players adjust their techniques according to the needs of the instrument as well as request instruments that suit their performance needs. For example, Banarasi tabla players are well known for striking their drums harder and playing more energetically than players from other traditions, and Banarasi tabla are similarly known for having thicker, more durable heads with large amounts of tuning paste that facilitate this style. The instrument design and performance style reinforce each other as beginners learn to strike harder while playing Banarasi tabla and then later require Banarasi tabla in order to do so.

38. In "Tabla Tuning on the Workshop Stage," I discuss at length an episode at Ekhlak's shop where a customer refused to pay him his standard rate and literally shoved a lesser amount of money into Ekhlak's shirt pocket before grabbing his instrument and running away.

Bibliography

Appadurai, Arjun. "Disjuncture and Difference in the Global Cultural Economy." *Public Culture* 2, no. 2 (March 1990): 1–24.

———. "Introduction: Commodities and the Politics of Value." In *The Social Life of Things: Commodities in a Cultural Perspective*, ed. Arjun Appadurai, pp. 3–63. Cambridge: Cambridge University Press, 1986.

Australian Quarantine and Inspection Service. "Import Case Details—Public Listing." http://www.theQuarantine.gov.au/icon32/asp/ex_casecontent.asp?intNodeId=8198614&intCommodityId=13334&Types=none&WhichQuery=Go+to+full+text&intSearch=1&LogSessionID=0 (accessed 3 April 2012, site discontinued).

Bates, Eliot. "The Social Life of Musical Instruments." *Ethnomusicology* 56, no. 3 (2012): 363–395.

Beck, Ulrich, Scott Lash, and Anthony Giddens. *Reflexive Modernization: Politics, Tradition and Aesthetics in the Modern Social Order*. Stanford, CA: Stanford University Press, 1994.

Bennett, Jane. *Vibrant Matter: A Political Ecology of Things*. Durham, NC: Duke University Press, 2009.

Bushberg, Jerold T. "Radiation Exposure and Contamination." Merck Manuals. Last modified June 2009. http://www.merckmanuals .com/professional/injuries_poisoning/radiation_exposure_and_contamination/radiation_exposure_and_contamination.html ?qt=&sc=&alt= (accessed 3 April 2012).

El-Ojeili, Chamsy, and Patrick Hayden. *Critical Theories of Globalization: An Introduction*. New York: Palgrave Macmillan, 2006.

Khan, Salman. "Anwar Tabla Maker." http://www.anwartablamaker.blogspot.com (accessed 17 October 2012).

Kippen, James. "The History of Tabla." In *Hindustani Music: Thirteenth to Twentieth Centuries*, ed. Joep Bor, pp. 459–478. Delhi: Manohar, 2010.

Kopytoff, Igor. "Cultural Biography of Things: Commoditization as Process." In *Social Life of Things*, ed. Arjun Appadurai, pp. 64–91. Cambridge: Cambridge University Press, 1986.

Latour, Bruno. *Reassembling the Social: Introduction to Actor Network Theory*. New York: Oxford University Press, 2005.

———. *We Have Never Been Modern*. Translated by Catherine Porter. Cambridge, MA: Harvard University Press, 1993.

Myers, Fred. "Wizards of Oz: Nation, State, and Production of Aboriginal Art." In *Empire of Things: Regimes of Value and Material Culture*, ed. Fred Myers, pp. 165–204. Santa Fe, NM: School of American Research Press, 2001.

reemixx. Post on tabla forum chandrakantha.com. Posted 11 August 2009. http://forums.chandrakantha.com/viewtopic.php?f=3&t =6306&hilit=Reemix#p37296 (accessed 14 October 2012).

Ritzer, George. *The Globalization of Nothing 2*. Thousand Oaks, CA: Pine Forge Press, 2007.

Roda, P. Allen. "Resounding Objects: Musical Materialities and the Making of Banarasi Tablas." Ph.D. diss., New York University, 2013.

———. "Tabla Tuning on the Workshop Stage: Toward a Materialist Musical Ethnography." *Ethnomusicology Forum* 23 (2014): 360–382, doi:10.1080/17411912.2014.919871.

World of Tabla. "Tabla Teachers in USA." http://world-of-tabla.com/tabla_teacher_usa.php (accessed 17 October 2012).

CHAPTER 7

Electric Turkish Coffee Makers
Capturing Authenticity for Global Markets

Harun Kaygan
*Assistant Professor
of Industrial Design*

*Middle East Technical University
Ankara, Turkey*

Whether and to what extent nations can still be considered the organizing principle of culture has been under close scrutiny for almost three decades.[1] As localities across the globe become more interconnected, cultural interaction defies and disrupts national boundaries.[2] The technologies of rapid communication and transportation, once considered to serve nation-building projects under technological nationalism,[3] today facilitate alternative transnational connections and global cultural formations.[4] In Arjun Appadurai's formulation, the global flow of people, ideas, capital, and technologies is no longer isomorphic—either contained within national borders or highly regulated by nation-states. It takes place more independently and is therefore patterned in increasingly dissimilar ways that challenge nation-states' efforts to channel and manipulate them.[5]

In the literature on globalization, these developments have been interpreted as an attack on cultural authenticity, including national cultural specificity, by global media and corporate practices. At one extreme end of the debate, globalization has been associated with homogenization, as it disseminates a global culture of Starbucks and Hollywood to worldwide localities.[6] Others have indicated a countermovement whereby global market practices make use of local cultures, as in food, fashion, and music, and assimilate those. In the process, it is not local difference per se that is eradicated but cultural specificity of the incorporated elements as they are stripped of their original meaning and significance. This argument found its most powerful expression in Jean Baudrillard's theory of a postmodern consumer culture that is based on simulation as depthless signification.[7]

Others have argued against this view, indicating the creative ways in which global imports—technologies, commodities, cultural forms—are appropriated at the receiving end. The term "indigenization" describes this process by which a local culture is constructed out of such imports and implies resistance to the imposition of a dominant culture via colonialism or global market practices. Bicycles, cars, and audio cassette players may be globally spread technologies, but in different localities they assume varying, even unique, uses and meanings.[8] In the field of material culture, Appadurai's work on Indian cricket and Daniel Miller's study of Trinidadian Christmas have provided influential examples as to how indigenized cultural elements are appropriated into not only local but national cultures, so much so that they are conceived as authentic expressions of national identity.[9]

Although significant in its objection to theories of homogenization and eventual cultural convergence, the major drawback of this perspective is that it foregrounds the construction of hybrid cultural forms and ignores the question of the maintenance of the traditional. It thus fails to account for the ways and contexts in which cultural authenticity in its contemporary transformations remain a concern for both producers and consumers.

In studies of cultural tourism, a similar discussion has engaged more directly with the concept of authenticity. Dean MacCannell famously suggested that tourism is the search for the authentic in an increasingly inauthentic modern world, which nevertheless can only provide a "staged authenticity," an empty representation. Later critique targeted not only the idea of stages but also the postmodernist view, such as that of Baudrillard, and revealed the limits of thinking with the concept of authenticity as an objective quality that can be preserved or lost. Instead, it has been argued that research should focus on the social construction of authenticity in specific tourism contexts and the related processes of negotiation and authentication.[10] Similarly, the way globally connected commercial practices make use of national cultural elements, that is, objects, practices, and ideas closely associated with a certain nation, cannot be assumed to rob them of cultural specificity and reduce them to superficial imagery. Inauthenticity of the outcome cannot be taken for granted; rather, processes of authentication, or deauthentication, need to be described in specific.

One step in this direction can be found in Michaela DeSoucey's study of "gastronationalism," defined as the two-way relationship between food-related practices and nationalist discourse within a global context. Specifically, she examines the policies of the European Union for designating certain food products as "cultural patrimony" of certain countries, which create monopolistic exceptions within the open-market structure of the union. In this case, gastronationalism appears in "a distinct organizational form . . . that prizes conceptions of tradition and authenticity as desirable rationales" for regulative action against the globalizing dynamics of the open market. In a similar vein, Sarah Bowen and Marie Sarita Gaytán built a case study of Mexican tequila on this concept to demonstrate how the nationalist discourse on heritage plays a part in the management of global commodity chains, namely, that of tequila.[11] Although both studies are important in underlining the relevance of authenticity for global commercial practices, they are still concerned with protectionist mechanisms and therefore understand references to cultural authenticity mainly in defensive terms. In other words, their cases place the nationalist arguments for the preservation of tradition *against* globalization effects—specifically, the homogenizing or hybridizing tendencies of the global market. In this, they mirror the theories of globalization I discussed above that depict national authenticity in disagreement with such tendencies and therefore risk a reductive understanding of a more complex process.

In this chapter, I argue instead that national cultural authenticity can be evoked not only to protectionist ends but also in a generative mode. In certain settings of technology and product development and design, authenticity can be an explicit discursive and practical concern, even coupled with discourses on national ownership and pride, yet this does not necessarily entail a traditionalist defense of existing traditions in the face of globalization. On the contrary, global markets, both as institutions and an idea, can provide the very thrust behind commercial technological projects that make use of nationally charged cultural material.

Actor-Network Theory Approach to Technological Objects

To investigate these complexities, close-up studies of such projects are indispensable. Technological objects are effective as cases to this end, for they can be used as nodes around which rich interactions between producers, engineers, designers, marketers, users, and nonhuman agents can be observed. From within these, discursive and practical repertoires can be extracted that shed light on the ways in which the local, national, and global are negotiated in everyday practices, around everyday technologies.

For this purpose, actor-network theory (ANT) provides a flexible yet potent framework. Accordingly, a technology should be analyzed as the product of the negotiations that take place during and after its development and design, as actors with varying interests are involved in the project. In ANT, the actors are not limited to the relevant professional or social groups; in fact, they are not even limited to humans: fuel cells, electrodes, and electrons; doors, keys, door closers, and human parts can be actors, as well.[12] Madeleine Akrich and Bruno Latour define "actant" as "whatever acts or shifts actions" and "actor" as "an actant endowed with a character," anthropomorphic or otherwise.[13] Any technology or technological object is constructed, and stabilized, through the articulation of such diverse actors in the form of a network, a "sociotechnical" assemblage.[14]

A key insight of ANT is that this coming together of heterogeneous entities shapes not only the final network (i.e., the technology) but also the entities themselves. The process whereby each actor is defined, attributed a character and a motivation, and enrolled in the project is called "translation."[15] In this, the actors are understood in a strictly performative mode. They are considered to have no inertia other than the day-to-day performances that describe them, represent them, translate them—social groups need group-making efforts; companies require being spoken for and about; products are held together with screws lest they fall apart.[16]

Foregrounding the enactments of technology within processes of development and design while also including considerations of the material agency of objects in such processes, ANT can be used to account for how authenticity is negotiated and objectified, in this case in the form of a technological object.

Researching Electric Turkish Coffee Makers

Kitchen appliances used to make Turkish coffee, namely, electric Turkish coffee makers, provide one such case. Electric appliances have been used to this end for at least a couple of decades in Turkey, with early examples being generic water heaters that were appropriated for preparing coffee. It was only in the mid-2000s that electric Turkish coffee makers emerged as a product category of their own with the first few commercially successful products that were designed specifically for this purpose. Having invested since 1995 in its commercially unsuccessful automatic coffee maker, Kahveset, Bayıner brought out a simpler product, Elektrikli Cezve, in 2002. The next year, Arzum started producing a similar appliance, Kahwe, and in 2005, its redesign made of metal, Cezve, was introduced. The latter product was designed by Kilit Taşı Tasarım, a design office that is well known in high-design circles, and became iconic of the product category.[17] Meanwhile, in 2004, Arçelik, a major white goods and electric appliances manufacturer and brand in Turkey, developed Telve, an automatic Turkish coffee maker (Figure 1). From 2006

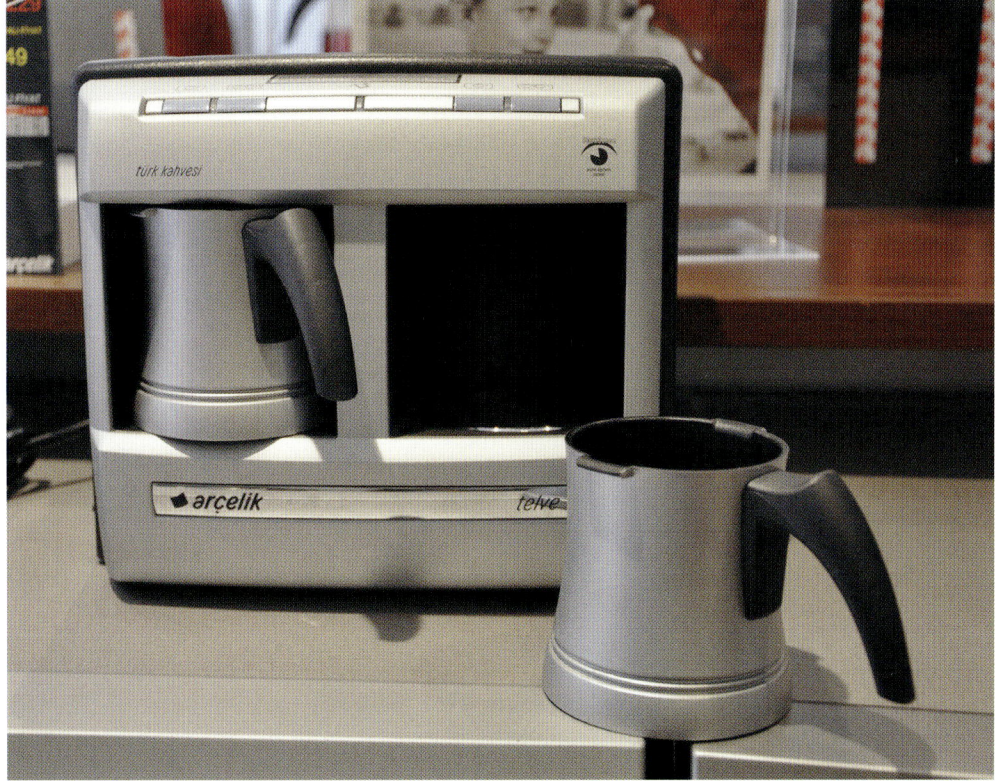

1.
Arçelik Telve, a well-received automatic Turkish coffee maker in the market since 2004. Photograph by the author.

on, various brands joined the competition. Between 2002 and 2010, the period I focus on in this study, around 27 products were launched, representing 13 brands.[18]

To research the technology development and design processes, I conducted interviews (in Turkish) with the designers, engineers, managers, and marketers (18 professionals in total) of 14 different electric Turkish coffee makers representing 9 different brands. The interviews were supported by the analysis of various objects (prototypes and coffee pots purchased by designers for research purposes) and documents (sketches, computer renderings, presentations showing design alternatives, etc.) produced during and after the processes.

Since the interviews revealed details of professional, and sometimes personal, relationships between and among participants and companies, confidentiality has been a concern. The fact that the participants gave information about relatively new products and technologies, as well as openly discussed their political views, accentuated the concern. In the analysis that follows, all participants and brand and model names are given pseudonyms both in the text and the notes.

A Liberal Neonationalist Project of Technology and Product Development

Turkish coffee is a hot, frothy coffee drink that is usually made in a metal coffee pot on the stove (Figure 2). Although there are different opinions regarding the best practice, it is generally prepared by putting coffee, sugar, and water inside the pot and heating the mixture until its froth starts to rise, then dividing it among the cups. It is then served with plenty of froth on the top and coffee particles inside the cup.

The oft-mentioned difficulty is that since the coffee pot needs to be taken off the stove before the coffee overbrims, it needs to be watched over in the duration of cooking. In response, the immediate solution has been to use electric heating to make the cooking process quicker and thus reduce the waiting time. Before the development of appliances specifically designed as Turkish coffee makers, plastic bare-element kettles were used for at least a couple of decades as a quick-and-dirty method to make coffee. Being inexpensive but unsafe due to the risk of electric shock, these were ultimately banned by Turkey's former Ministry of Industry and Trade in 2008.[19] Another suggested solution has been to fully automate the process and to dispense with the requirement of watching over the pot. A variety of different approaches has produced a number of automatic Turkish coffee makers since 1995.

Indeed, the executives I interviewed pointed out that providing solutions to this difficulty has been the major priority in the electric Turkish coffee maker projects. In general, they described a need to improve Turkish coffee making, which in turn offered an opportunity to develop a technological product. For instance, Naim Keskin, the CEO of Sevilen, which is a major home appliances company in Turkey, defined the very mission of his company as making the housewife's tasks in the kitchen easier and quicker by providing them with what he described as "minutae-products for which they'd say 'Oh, if there were such a thing, it'd be so much easier

2.
Two Turkish coffee pots, which are typically used to prepare Turkish coffee on the stove, in a kitchen. Photograph by the author.

for me!'" His examples included preparing "Turkish tea" and "Turkish coffee," for which Sevilen already had products in the market at the time of the interview, as well as others, such as making stuffed vine leaves or preparing *mantı* (a pasta dish that is often translated as "Turkish ravioli"), which they had "not yet been able to design a product for."[20]

As evident from the common adjective, Turkish, that qualifies the mentioned food items, there is more to these projects than commoditization of the vernacular, as that which is commoditized is considered to be national. Selim Cansu, the executive of Özkar, another manufacturer specializing in kitchen appliances, discussed Turkish coffee explicitly as a national tradition:

> In the world, there are three or four different types of coffee, . . . but Turkish coffee has a unique preparation technique. In Turkey, they make it relatively stronger. In the south, especially in Syria, Israel, Lebanon, Arabia, it's even stronger. They call it "*mırra,*" they make it very strong, stronger than ours. In Greece, too, it's more or less the same style of coffee as ours; theirs is a bit lighter. . . . Within this group that I tell you, it's the same preparation technique. I suppose it's the Ottoman influence. The preparation technique has spread there.[21]

The executives, engineers, and designers I interviewed acknowledged that what is called "Turkish coffee" is consumed outside Turkey, too. The research activities of one product development team included visits to Lebanon and Greece to document relevant practices, whereas another project was instigated by observations that "Arabs" knew about Turkish coffee and thus represented a potential market.[22] Similarly, Aydın Gürcan, a freelance product designer who designed electric Turkish coffee makers for a number of manufacturers, mentioned that one of the concepts they developed had been inspired by what he considered a Greek practice to cook coffee on embers buried in sand.[23] The dispersion of coffee practices beyond national boundaries has not, however, made the actors contradict the Turkishness of Turkish coffee. Instead, the national character of the practice was defined in temporal rather than spatial terms. Accordingly, it is a strictly Turkish tradition with its origins in the Turkish national past, being historically based in the so-called Ottoman culture. The more extensive distribution of coffee practices is thus explained away as being due to the historical Ottoman influence on the larger geography.

Selim Cansu continued his discussion by describing how and why they worked on a Turkish coffee maker:

> Now here we're simply making this [technique] a bit easier. Espresso, cappuccino have made it easier, mechanized and spread it worldwide. If we mechanize it, too, our machine will also be spread worldwide. . . . If we can come up with a proper machine, we can become another Italy. You see, Italy dominates the global market with espresso. I mean, with our Turkish coffee machine, we, too, can make a serious impact.

The excerpt is significant in that it draws attention to a number of key aspects as to where the project stands with regard to gastronationalism. First of all, the comparison with "Italian coffee" was a common motif in executives', engineers', and designers' accounts. If there exists an Italian coffee maker, by extrapolation, there should be an electric Turkish coffee maker, too. In this manner the comparison dictates a classification system in which food practices in general, and coffees in particular, are sorted by nations.[24] Discursively, this is achieved by "we," which denotes the executive's company, electric appliance producers in Turkey, and the Turkish nation interchangeably. In effect, it amounts to a series of nationalizations: a geographically dispersed practice of coffee making is translated into a national tradition, the electric Turkish coffee maker project becomes a national project to mechanize that tradition, and the company recasts itself as a national actor that will undertake this task.

Before proceeding any further, it is important to emphasize the term "nationalize" here. What it means is that actors, material objects, and practices that are associated with a certain nation cannot be assumed to be nationally bound by themselves but are enacted as such, that is, nationalized, in discourse and in practice. Assuming otherwise amounts to an oversight that has been widespread in the literature on nations and nationalisms. For instance, political geographer John Agnew called "the territorial trap" the misconception that nation-states are fixed units that contain well-bound cultures. Michael Billig pointed out the existence of a "sociological common sense" that takes nations as *the* legitimate unit of academic research, whereas Wimmer and Schiller later termed this tendency "methodological nationalism." Similarly, Rogers Brubaker indicated the "substantialist" underpinnings of much research in theories of nationalism, in which nations as they are described by nationalist politics are taken up as theoretical categories.[25] All in all, to assume the existence of nationally bound cultures and to take them as unproblematic objects of analysis has a reifying and even naturalizing effect on what is posited as national culture by the actors.

This further adds to the importance of close-up studies in which practices of nationalization can be investigated in detail. It is particularly crucial to show how such practices are articulated to actors' declared interests and expectations of benefit. The above excerpt makes a very lucid association in this regard between the national ownership of Turkish coffee (despite its spread to other countries under Ottoman rule) and the potential benefits of its commoditization for Turkey and for the company as its representative.

As implied by the representative status Cansu ascribes to himself and his company, nationalization is profitable not only in economic terms but also symbolically. In their publicity, producers have routinely referred to their contributions to what they consider the national project to promote Turkish coffee globally. One brand declares on its website that its mission is to "introduce Turkish culture to the world" via its electric coffee makers. A company representative from a café chain, which also sells its own Turkish coffee makers, claimed in a publicity interview that they aim "to bring Turkish culture of coffee to the place it deserves" and added, "We will be proud when we see Turkish coffee in restaurant menus worldwide."[26] Likewise, two executives told me proudly how they "taught Turkish coffee" to the Chinese manufacturers they worked with.[27]

Such pride and responsibility is decidedly nationalist. Particularly, it aligns with "liberal neonationalism," a term coined by Tanıl Bora in his categorization of Turkish nationalisms. According to Bora, liberal neonationalism matured in the late 1980s with the ongoing integration of Turkey into the global economy. At the discursive level, it has mainly adopted modernist aspirations to "catch up with the West" and is characterized by attempts—both in discourse and in practice—to prove, repeatedly, that Turkey has already accomplished this. Accordingly, national pride is gathered less out of a claim to the uniqueness of Turkish people than through comparisons with the so-called developed countries, especially regarding the extent of articulation to the global culture of consumption[28]—which indubitably includes the ability of national industries to develop globally competitive technologies, products, and brands. In the context of electric Turkish coffee makers, it culminates in the argument that if Turkey is on a par with other nations, it should be able to exploit—commoditize and market—its culture as does Italy.

The issue is then, to return one last time to the above excerpt from Cansu, "to come up with a proper machine" for this task. But how is propriety to be assessed? As Keskin was narrating to me their concerns in the product development process, he noted that when they had started the project, there had already been some coffee makers on the market, albeit in need of much improvement. This was with reference to both the earliest makeshift coffee maker kettles and some of the newer electric Turkish coffee makers. Even though he cited problems of safety and usability, his understanding of propriety was not limited to internal criteria, such as whether the product is fail proof, well designed, or commercially successful. More importantly, they were not products "that you can sell with pride in London, in England, in the US, in New York, wherever."[29] So propriety regards the extent to which the products can be marketed on equal terms with those of the Western countries. Hence, technology becomes a bearer of nationalist pride. The otherwise technical issues of safety and usability become subordinated to the mission to prove the technological adequacy of Turkey by developing electric Turkish coffee makers that are "proper" for global markets.

I heard a similar complaint from Nilgün Bircan, a freelance designer who had been part of a team that had designed a Turkish coffee maker for the company Özkar. She explained to me how they felt when they found out that it was only a small manufacturer that did research and development work on coffee machines:

> I felt sorry that these people do this whilst [big companies] don't. [Big companies] have always been timid when it came to this matter. . . . [The manager of a small manufacturer] has invested, I don't know, one million dollar in this, he may lose it, but he takes the risk. . . . If you don't do this as big companies, this sort of people will.[30]

In Bircan's argument, the *aspiration* to represent Turkey with a proper Turkish coffee machine is further transformed into the *responsibility* of Turkish companies to do so. In general, this is conveyed as a distaste for "poorly" designed products and brands. As demonstrated by William Mazzarella in his study of advertising in India, a postcolonial sense of technological and cultural inadequacy can give rise to a fear of looking "provincial" when compared to "world-class" brands. What this implies is that in the global market, cultural specificity can be accepted only in a certain form of presentation, which is perceived by the actors in production and design as a standard to reach.[31] In this case, the standard is technological; it dictates a technological form for the national cultural content.

Overall, the arguments indicate that this is a specific manifestation of technological nationalism as articulated to liberal neonationalist discourse, so that it is not oriented toward nation building in the strict sense of the term but nurtures a conception of national pride and responsibility via the establishment of technological prowess. The expectation of propriety carries the question further into the actual design process, in which the actors manage national cultural authenticity on the one hand and adequacy for global competitiveness on the other. In what follows, I will describe a series of instances where either one or both of these requirements directed the design of electric Turkish coffee makers. My aim is to show the extent to which authenticity mattered in the process, from the enrollment of designers, through the technology development process, to the authentication of the design decisions.

Enrolling Designers

As I described above, understanding design as a sociotechnical setting requires an analysis of the processes of network building that take place at that setting, as various actors are defined and enrolled in the project. In most cases I researched, the process started with producers recruiting designers. And even at this early step, nationality appeared as a criterion for enrollment.

Describing their choice of designers, Naim Keskin told me that they first thought of contracting one of the Italian designers they used to work with but changed their minds. They were concerned that whereas a Turkish designer would know how to design a Turkish coffee maker, others would not. Later, Keskin came across a modernized shisha design in an exhibition in Istanbul and contacted its designer, Fuat Tuncay.[32] When I interviewed Tuncay himself, he speculated that it was his "obsession with cultural things like shisha and so on" that had made the executive offer him the project.[33]

A similar story was offered to me by Vedat Semerci, who designed an electric Turkish coffee maker for Atasan, an OEM company specializing in plastics. For design services, Atasan normally worked with what he called "an English company" producing electrical parts, but they were not satisfied with the design alternatives they received for their electric Turkish coffee makers:

> Although this worked with toasters and kettles, when it came to electric coffee pots. . . . Since the English didn't know Turkish coffee and the way it's cooked, . . . they designed things like small plastic kettles, only with a handle. That's why [the company] found me.[34]

In both stories Turkish designers are contrasted with designers of other nationalities, drawing attention to some asymmetry in knowledge. This amounts to a nationalization, as designers are translated to Turkish nationals and enrolled on that basis. In the second story, the comparison is mirrored in material objects. There is a difference between Turkish coffee makers and other kitchen appliances, such as kettles, which lack the cultural specificity. Therefore, what the designer later called the "standard" ("*klasik*") design practice is not sufficient for the project and needs to be somehow complemented by the cultural background of an insider. Similarly, in the first story, the perceived cultural specificity of shisha and Turkish coffee bring them together as "cultural things" and is potentially influential in the designer's enrollment.

In general, companies seem to have considered it important to keep the designs authentic to traditions. Just as the Italian coffee maker was considered a successful technological transformation and commoditization, kettles represented ubiquitous, "standard" designs that need to be avoided in national projects. Therefore, they relied on those designers with insider knowledge of and interest in national traditions to make the correct decisions.

That the Turkish have ownership of, even a monopoly on, the knowledge and skills related to Turkish coffee was a general theme that I found throughout my research. However, this was not the only way in which nationality mattered in stories of designers' enrollment. The designer, Bircan, told me a third story that featured a more complex network of relationships. It all started when Bircan and her colleagues participated in an international exhibition in Italy on the subject of Mediterranean cultures and design. In addition to their own conceptual designs on Turkish food habits, the organizers asked them to bring, in her words, "something that promotes something from your culture." So they decided to make an exhibit on "Turkish tea culture." For this, they contacted Özkar, a manufacturer of electric tea makers, to borrow one of their designs. A year after the event, this time the manufacturer contacted them for its new electric Turkish coffee maker project, drawing on their former correspondence regarding tea makers.[35]

Although resembling the first story above, Bircan's narrative involves three actors: the designers, the organizers, and the company. The exhibition links the manufacturer and the designers, and in turn, the designers link the manufacturer with the international design community by showcasing its product. Arguably, both relationships are prompted by the organizer's interest in Turkish culture as one identity within the Mediterranean region. In effect, this nationalized the designers, the company, and their product in one sweeping gesture. This indicates, above all, that such international connections can motivate the employment of national categories in discourse and in practice. Since the Great Exhibition of 1851, international exhibitions have had such a role in imposing a perspective of international competition not only within the so-called developed nations but onto peripheries, presenting nationhood as the primary criterion for classifying products, visual styles, and design institutions.[36] The story can be considered a small-scale instance of such an imposition, which was well received by the designers and the manufacturer.

Black-Boxing the Technique Qua Tradition

Further into the design process, I found that concerns with national cultural authenticity were diffused in the actual practices of product development and design. To start with, since the Turkishness of Turkish coffee was defined predominantly in temporal terms, that is, as a national tradition with its roots in the Ottoman past, research, development, and design teams concentrated on identifying the one correct coffee-making practice that has traveled from that past to this day. For this purpose, the teams engaged in a variety of research activities. I already mentioned above that one of the teams went on field trips to examine related coffee traditions, including visits abroad. Others observed and interviewed coffee masters in restaurants and consulted Turkish coffee brands. Most teams also carried out historical research. Some studied books on the culture and history of coffee, as well as on the more general topic of Turkish traditions. Many purchased sets of old Turkish coffee pots to study which cooking practices each affords (e.g., holding in a certain manner, putting on fire or on a brazier, etc.).[37]

The coffee-making practices established through research were then objectified in a variety of electric Turkish coffee makers through several strategies, for instance, form giving, product specification, and technical detailing. To illustrate the way these practices are captured in the materiality of a technological object, I will discuss one project, Urteknik's Dibek automatic Turkish coffee maker, in which the strategy was mainly to develop a series of new technologies to automate the practice.

In the Dibek project, the concentrated focus on automation made the designers and engineers attribute much significance to the technology rather than on the product in general. One engineer, Nihat Gözcü, described their principal achievement as "the presentation of the technique which describes the ideal way of cooking Turkish coffee by means of an ideal machine," one on which a variety of products can then be based.[38] Consequently, there was an unmistakable emphasis in their accounts on the technique as *the* unique aspect of Turkish coffee. For example, Esat Soylu, another engineer in Urteknik, claimed that

> Turkish coffee is not about a quality of the coffee itself. It's about the technique. . . . Nor is it the aroma. . . . Change the espresso machine as much as you like, you can't make a Turkish coffee machine out of it. You can't get this taste, this texture. You can't make a bulb out of a candle.

Any consideration, such as achieving proper taste or amount of froth, is deemed contingent on the correct description of the technique. What the product, and the technologies used in it, must accomplish is to replicate what the human being does via electromechanical means, as Soylu noted, "just as if it were a robot with a coffee pot."[39]

However, this technique is not simply a series of abstract steps but highly charged as a national tradition. For instance, by repeating the comparison with espresso, the Italian coffee, Soylu alludes to the status of Turkish coffee as a Turkish tradition. As frequent is the use of "we" to underline the collective knowledge of the cooking practice, such as when Gözcü suggested that what the final product reproduces is "how *we* do it when *we* do it in the coffee pot."[40] In this respect, what is being replicated is not so much a technique as a technique qua tradition, which is handed down through generations.

What then is this technique, and how was it replicated in the final product? The research and development team studied the problem extensively, including questionnaires, tours, focus groups, and cooking tests with the aim of isolating an abstracted technique and subsequently mechanizing it. The following is a description, as narrated by my participants,[41] of the way the automated Dibek makes Turkish coffee, with comparisons to manual practice (Figure 3).

As the first step, the user puts ground coffee and sugar in the product's container, places the container in its slot, and presses the button. The machine pumps water from its tank, creating a vortex, whose counterpart in the traditional method is to stir with the spoon. In the second step, a person following the manual practice would place the coffee pot on a heat source, most typically the stove. Inside the machine, this is achieved via a rising platform that lifts the heater so that it touches the bottom part of the container. The movement literally replicates the coffee pot's transfer onto the stove. The moveable heating mechanism is crucial, not at this step, but after cooking ends, when the heater once again moves down away from the container. This is because boiling needs to stop at the exact moment coffee froth reaches sufficient height to avoid overflowing and to achieve perfect froth. In the third step, the person would watch over the coffee and wait for the froth to rise. Inside the machine, this action is delegated to a sensor, which "watches the froth develop . . . hundreds of times each second."[42] In the fourth step, the person would decide that the froth has risen sufficiently, that is, reached the brim, and take the coffee pot off the stove.

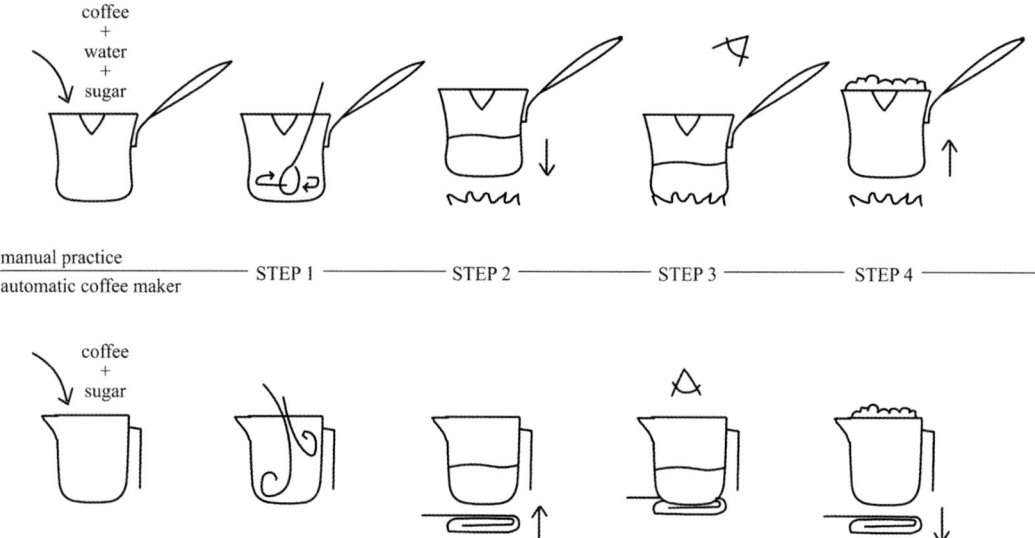

3.
A visual depiction of the participants' comparison, step by step, between the manual Turkish coffee-making technique and the technology they developed. Illustration by the author.

Inside the machine, it is the processor that makes the decision and activates the moving platform to remove the heater from the container, mimicking the user taking the coffee pot off the stove. Last, when the coffee is ready, an alarm sounds, and the user can take out the container to pour the coffee into cups.

From the first step onward, the process of abstraction and delegation is evident, whereby the observed practice is divided into steps and associated functions, each of which is delegated to an actor. Effectively, what the design team achieved is a rearrangement of the actors in the sociotechnical assemblage that makes Turkish coffee. Replacing the pot on the stove with an automatic alternative, the new design redistributes the agencies, that is, roles and skills, among various actors in the kitchen, including the user, coffee, sensors, water jets, etc. In the process some of these are made obsolete, such as spoons and stoves, whereas others remain as they are, such as the role of coffee particles in water. A significant part of the user's coffee-making skills and experience is also made redundant, being disembodied and distributed to a number of actors. Stirring is delegated to the water jet, watching to the sensors, and the user's judgment to the processor. Strictly speaking, the processor is not a decision maker but an objectification of the decisions made by engineers in the past. The decisions had already been made during the design process and black-boxed[43] in the product. A major implication is that the new technology defines what steps (e.g., stirring, watching the froth, dividing coffee into cups) the traditional technique for making Turkish coffee comprises and what the outcome should be like (frothiness, consistency, etc.). Furthermore, these are established and inscribed in the product irrevocably, so that the object becomes the carrier of the engineers' and designers' inscriptions to future settings.[44]

Enrolling Users, Delegating Authenticity

Still, the success of the technology requires that users subscribe to their place in it. For this, they have to consent to their deskilling, as well as to their newly prescribed roles regarding the operation and cleaning of the machine. In this regard, one strategy to achieve the enrollment of users was via projected users, that is, future users as imagined and invoked by the professionals involved in product development. In the interviews, designers and engineers speculated freely about the capabilities and expectations of users and explained their decisions with reference to them. As a second strategy, certain actors can be designated to represent the future real user. This role was often left to company workers, mainly in drinking tests within the company.[45]

I encountered systematic user tests only in the design of Dibek. At the very beginning of the project, focus groups were conducted with potential users to document their views as to whether Turkish coffee can be made automatically, that is, by a machine. The opinion was largely negative, with many users citing the complexity of the preparation method.[46] Once the first prototype was complete, another set of focus groups was organized. The product's designer, Tuna, narrated to me that the tests started with asking the participants how they made coffee. They then described their practices, often making divergent suggestions, and debated among themselves which practice was preferable. Afterward, however, during the blind taste test,

> around 90 percent favored the coffee from our machine. So, umm, it's interesting that all these people who describe coffee differently liked the same coffee. . . . [The different practices the users described] are then like, umm, how do I put it, they're like myths then.[47]

Essentially, what the blind taste test does is to posit the represented user's taste as the definitive criterion in evaluating whether they consent to their prescribed role. If they like the taste of the coffee, they are considered to have subscribed to the technology; by extrapolation, the same is assumed for the future real user, too. The key point is that enrollment is ensured by defining the user solely in terms of their gustatory expectations—excluding their past experience, practical knowledge, the experiential aspects of coffee making, etc.—so that their voiced opinion against automatic Turkish coffee machines can be circumvented. This made it possible for the designers and engineers to dismiss users' alternative definitions of traditional coffee-making practice as "mythical" or "psychological."[48] In contrast, the propriety of the technique that was abstracted and applied by the engineers was sanctioned.

Furthermore, according to the designers, in the user tests the technology was "proven" to cook not only as good as manual practice but better. Erhan Baygun, a product designer in Urteknik, interpreted the results as follows:

> When the coffee is made manually, sometimes it has froth, sometimes it's all over the place, sometimes something else. But the coffee from the machine is usually consistent in both taste and froth. That's because handmade coffee is incidental to the maker's skills.[49]

The implication is that when Turkish coffee is defined as technique qua tradition and the user is appointed as the deskilled judge of traditional taste, the machine appears to make coffee by itself. The outcome is no longer "incidental to the maker's skills." Therefore, the deskilling of the user who switches from the coffee pot to the coffee machine makes the Turkish coffee-making practice available to those who had never had the necessary skills in the first place. Following the discourse on the national ownership of Turkish coffee knowledge and skills, Tuna argued that this includes the "foreigners":

> They will be able to make [coffee] now. Previously they couldn't. I mean, give the coffee pot
> to a foreigner, what are they going to do with it? They wouldn't know how to make coffee!
> This is why we say it's something that comes to us via generations.[50]

Put differently, it is not only the skills of the Turkish user but that which "comes to us via generations"—that is, authenticity itself—that is delegated to the machine. The automatic Turkish coffee machine is constructed as the objectification of the technique qua tradition.

Perhaps not surprisingly, this argument did not elicit a positive response from the designers of other Turkish coffee makers, especially from those who defined traditional practice in a broader manner than an abstracted technique.[51] But according to the team involved in the technology development for Dibek, the deskilling did not make the new technology, or Turkish coffee in general, less authentic or less national by obviating the traditional practice. On the contrary, the authentic essence of Turkish coffee is extracted as a technique, black-boxed in the form of a technology, and commoditized as a product so that it becomes comparable to other globally spread technologies of coffee making such as espresso machines.

Concluding Discussion

The aim of this chapter was to counter the arguments in the literature on globalization that it makes national cultural authenticity in its traditional nationalist form obsolete and replaces it with either postmodern or hybrid cultural forms. On the contrary, as I have shown, commercial practices of design and technology development still engage with authenticity, not only from a defensive position that favors protectionist policies but also in a generative mode in which vernacular cultural elements, conceived as national cultures, are elaborated to develop new technologies and commodities. Nor is the final product perceived by the product development teams to be a hybrid in the sense that the teams brought together global and local concerns, elements, technologies, etc.; it is enacted as a pure expression of national cultural authenticity. For the teams, this view does not contradict the fact that the projects were globally connected from the very beginning, that is, involved global actors, conceived in comparison with "other coffees," and targeted propriety for global markets. On the contrary, the very success of the machine seems to have depended on the essence of the tradition captured in it.

It is important to note that "generative" does not mean "positive" or "benign." Writing in 1992, Hugh Aldersley-Williams argued that despite globalization, nations remain "the principal designator of cultural character" owing to "the facts of political, economic and commercial life." He then called for a design practice that engages more actively with national cultures—one that could foster "benign new nationalisms," which "may no longer serve much political purpose, but [which] could contribute materially to company performance." In this manner, he argued, "design could begin to restore to artefacts some of the meaning they have lost as societies became more secular, more industrialized, and more intertwined."[52] More recently, his arguments have been challenged by Viviana Narotzky, who indicated that what Aldersley-Williams advocated was a "commodity-led, business-oriented formalism," which disregards the reductive, even stereotyping, way in which national characteristics have been handled by commercial practices.[53] Her arguments parallel earlier accounts in design historical research that demonstrate the reductiveness of such formalistic approaches.[54]

I contend that there is more to criticize in the suggestion that a globally situated and commercially motivated nationalism is without political implications. The case study of electric Turkish coffee makers shows that the very practice of commoditizing national cultures is closely related to a distinct form of Turkish nationalist discourse. The liberal neonationalist discourse, coupled with technological nationalism, becomes the basis for nationalist pride and responsibility in commercial dealings within, or with an eye to, global markets. Technology and product development practices regularly resort to discourses of national ownership and national cultural continuity. In so doing, these practices have the effect of nationalizing—that is, rendering national—those vernacular material cultures and practices that are not necessarily homogeneous within or limited to national borders. They posit as they employ nationhood. Captured in the materiality of the final product, as it is spread to kitchens around Turkey—and, if the technologists' ambitions are satisfied, all around the world—an otherwise richly variant practice of Turkish coffee making risks being reduced to a cultural package promoting a single abstracted technique.

The case of Turkish coffee provokes further questions as it provides insights regarding the automatization of national traditions at the intersection of global markets, nationalist discourses, and vernacular everyday practices. One of these regards the changing faces of technological nationalism in response to the intensified globalization effects: Today's technological nationalism is not concerned so much with radios and railroads that make nations possible as with kitchen appliances that represent authentic national cultures competing in a global market. A second question is that of domestication of technologies that are themselves derived from practices in the domestic sphere and developed for global marketability. To respond to this question, it seems essential to look at real users' responses to, and their myriad creative appropriations of, those technologies in everyday life. Use is never simply a matter of assessing a product's success in the market or its rate of acceptance in households but an indispensable part of the sociotechnical existence of the Turkish coffee maker. Overall, to fully account for the capturing and commodification of authenticity in the form of technical objects, it will be necessary to extend the field of investigation and not to be contented with aggregative explanations that presume the homogenization or hybridization of cultures across the globe.

Notes

1. Mike Featherstone, "Global Culture: An Introduction," in *Global Culture: Nationalism, Globalization and Modernity*, ed. Mike Featherstone (London: Sage, 1990), 1–2; Featherstone, "Islam Encountering Globalization: An Introduction," in *Islam Encountering Globalization*, ed. Ali Mohammadi (Oxon: RoutledgeCurzon, 2002), 1–13.

2. Kenichi Ohmae, *Borderless World* (London: Collins, 1990); Ulf Hannerz, "Cosmopolitans and Locals in World Culture," in Featherstone, *Global Culture*, 237–252.

3. Maurice Charland, "Technological Nationalism," *Canadian Journal of Political and Social Theory* 10, nos. 1–2 (1986): 196–220. For a more extensive review of the phenomenon, see David Edgerton, *The Shock of the Old: Technology and Global History Since 1900* (London: Profile Books, 2008), chap. 5. For a discussion of nationalism as cultural unification and homogenization, see Benedict Anderson, *Imagined Communities*, rev. ed. (London: Verso, 2006); Ernest Gellner, *Nations and Nationalism* (Oxford: Blackwell, 1983); Anthony D. Smith, *Nationalism and Modernism: A Critical Survey of Recent Theories of Nations and Nationalism* (London: Routledge, 1998).

4. Peter Lyth and Helmuth Trischler, "Globalisation, History, and Technology: An Introduction," in *Wiring Prometheus: Globalisation, History, and Technology*, ed. Peter Lyth and Helmuth Trischler (Aarhus: Aarhus University Press, 2004), 7–20; see also Peter J. Hugill, *Global Communications Since 1844: Geopolitics and Technology* (Baltimore: Johns Hopkins University Press, 1999); Vaclav Smil, *Prime Movers of Globalization: The History and Impact of Diesel Engines and Gas Turbines* (Cambridge, MA: MIT Press, 2010).

5. Arjun Appadurai, "Disjuncture and Difference in the Global Cultural Economy," in *Modernity at Large: Cultural Dimensions of Globalization* (Minneapolis: University of Minnesota Press, 1996), 27–47; Arjun Appadurai, Ashish Chadha, Ian Hodder, Trinity Jackman, and Chris Witmore, "The Globalization of Archaeology and Heritage: A Discussion with Arjun Appadurai," *Journal of Social Archaeology* 1, no. 1 (2001): 35–49.

6. Cees J. Hamelink, *Cultural Autonomy in Global Communication* (London: Longman, 1983); Herbert Schiller, *Communication and Cultural Domination* (New York: M. E. Sharpe, 1976).

7. Jean Baudrillard, *Simulations* (New York: Semiotext(e), 1983); Theodore Levitt, *The Marketing Imagination* (London: Collier-Macmillan, 1983); David Morley and Kevin Robins, *Spaces of Identity: Global Media, Electronic Landscapes and Cultural Boundaries* (London: Routledge, 1995), 109–118.

8. Tiina Männistö-Funk, "The Crossroads of Technology and Tradition: Vernacular Bicycles in Rural Finland, 1880–1910," *Technology and Culture* 52, no. 4 (2011): 733–756; Viviana Narotzky, "Our Cars in Havana," in *Autopia*, ed. Peter Wollen and Joe Kerr (London: Reaktion Books, 2002), 169–176; Andre Millard, "Audio Cassette Culture and Globalisation," in Lyth and Trischler, *Wiring Prometheus*, 235–250.

9. Arjun Appadurai, "Playing with Modernity: The Decolonization of Indian Cricket," in Appadurai, *Modernity at Large*, 89–113; Daniel Miller, *Modernity: An Ethnographic Approach; Dualism and Mass-Consumption in Trinidad* (Oxford: Berg, 1994), 318–321.

10. Dean MacCannell, *The Tourist: A New Theory of the Leisure Class* (Berkeley: University of California Press, 1999), 2–3, 94. For the critique, see Edward M. Bruner, "Abraham Lincoln as Authentic Reproduction: A Critique of Postmodernism," *American Anthropologist* 96, no. 2 (1994): 399–401; Eric Cohen, "Authenticity and Commoditization in Tourism," *Annals of Tourism Research* 15, no. 3 (1988): 374. For an example for the way authenticity is constructed differently in a range of settings, see Bruner, "The Maasai and the Lion King: Authenticity, Nationalism, and Globalization in African Tourism," *American Ethnologist* 28, no. 4 (2001): 881–908.

11. Michaela DeSoucey, "Gastronationalism: Food Traditions and Authenticity Politics in the European Union," *American Sociological Review* 75, no. 3 (2010): 433; Sarah Bowen and Marie Sarita Gaytán, "The Paradox of Protection: National Identity, Global Commodity Chains, and the Tequila Industry," *Social Problems* 59, no. 1 (2012): 70–93. For the relationship between national cuisines and nationalism in general, see David Bell and Gill Valentine, *Consuming Geographies: We Are Where We Eat* (London: Routledge, 1997).

12. Michel Callon, "The Sociology of an Actor-Network: The Case of the Electric Vehicle," in *Mapping the Dynamics of Science and Technology*, ed. Michel Callon, John Law, and Arie Rip (London: Macmillan Press, 1986), 19–34; Bruno Latour, "Where Are the Missing Masses? The Sociology of a Few Mundane Artifacts," in *Shaping Technology/Building Society: Studies in Sociotechnical Change*, ed. Wiebe E. Bijker and John Law (Cambridge, MA: MIT Press, 1992), 225–258.

13. Madeline Akrich and Bruno Latour, "A Summary of a Convenient Vocabulary for the Semiotics of Human and Nonhuman Assemblies," in Bijker and Law, *Shaping Technology/Building Society*, 259.

14. For the term "sociotechnical," see Thomas P. Hughes, *Networks of Power: Electrification in Western Society, 1880–1930* (Baltimore: John Hopkins University Press, 1983), 465.

15. Michel Callon, "Some Elements of a Sociology of Translation: Domestication of the Scallops and the Fishermen of St Brieuc Bay," in *Power, Action and Belief: A New Sociology of Knowledge?*, ed. John Law (London: Routledge & Kegan Paul, 1986), 196–223; Callon, "The Sociology of an Actor-Network," 20–26.

16. Bruno Latour, *Reassembling the Social: An Introduction to Actor-Network-Theory* (Oxford: Clarendon, 2005).

17. For an illustrative magazine article on the success of the product, see Fadime Çoban Bazzal, "En Başarılı 20 Yenilikçi Ürün," *Capital*, March 2007, 124–127.

18. It is difficult to give a reliable final number due to the relatively short lifespan of some products. Many companies have bought OEM products from manufacturers based in Turkey or China and marketed them for short periods, often before switching to their own designs. This is further complicated by the use of shared, or highly similar, designs by more than one brand, where the count is affected by different opinions as to what constitutes a unique design.

19. See "Plastik Kahve Pişiriciler Toplatılıyor," *Zaman*, 20 July 2008, http://www.zaman.com.tr/haber.do?haberno=716142 (accessed 10 December 2012).

20. Naim Keskin (CEO, Sevilen), interview with the author, 23 December 2009. All interview excerpts are translated from Turkish by the author.

21. Selim Cansu (CEO, Özkar), interview with the author, 29 July 2010.

22. Kerim Tuna (product designer, Urteknik), interview with the author, 22 December 2009; Cevdet Ata (production engineer, Sümer), interview with the author, 2 July 2010.

23. Aydın Gürcan (product designer and CEO, AG Design), interview with the author, 21 June 2010.

24. Bell and Valentine, *Consuming Geographies*, 165.

25. John Agnew, "The Territorial Trap: The Geographical Assumptions of International Relations Theory," *Review of International Political Economy* 1, no. 1 (1994): 53–80; Michael Billig, *Banal Nationalism* (London: Sage, 1995), 51–52; Andreas Wimmer and Nina Glick Schiller, "Methodological Nationalism and Beyond: Nation-State Building, Migration and the Social Sciences," *Global Networks* 2, no. 4 (2002): 301–334; Rogers Brubaker, "Rethinking Nationhood: Nation as Institutionalized Form, Practical Category, Contingent Event," *Contention* 4, no. 1 (1994): 5.

26. Arzum, "Arzum: From Past to Future," http://arzum.com.tr/en/company/ (accessed 10 November 2011); Eda Terçin, "Ödüller Kalitemizi, Üretim Gücümüzü Ortaya Koyuyor," *Capital Online*, 1 March 2011, http://www.capital.com.tr/oduller-kalitemiziuretim-gucumuzu-ortaya-koyuyor-haberler/22797.aspx (accessed 8 January 2012).

27. Keskin, interview; Cansu, interview.

28. Tanıl Bora, "Nationalist Discourses in Turkey," *South Atlantic Quarterly* 102, no. 2 (2003): 440–445. For a more extended discussion in Turkish, see Bora, *Medeniyet Kaybı: Milliyetçilik ve Faşizm Üzerine Yazılar* (Istanbul: Birikim Yayınları, 2006).

29. Keskin, interview.

30. Nilgün Bircan (freelance designer), interview with the author, 28 June 2010.

31. William Mazzarella, "'Very Bombay': Contending with the Global in an Indian Advertising Agency," *Cultural Anthropology* 18, no. 1 (2003): 41; Derya Özkan and Robert J. Foster, "Consumer Citizenship, Nationalism, and Neoliberal Globalization in Turkey: The Advertising Launch of Cola Turka," *Advertising and Society Review* 6, no. 3 (2005): doi:10.1353/asr.2006.0001.

32. Keskin, interview.

33. Fuat Tuncay (product designer and CEO, Tuncay Design), interview with the author, 22 December 2009.

34. Vedat Semerci (freelance product designer), interview with the author, 30 June 2010.

35. Bircan, interview.

36. Abigail S. McGowan, "'All That Is Rare, Characteristic or Beautiful': Design and the Defense of Tradition in Colonial India, 1851–1903," *Journal of Material Culture* 10, no. 3 (2005): 263–287; Artemis Yagou, "Facing the West: Greece in the Great Exhibition of 1851," *Design Issues* 19, no. 4 (2003): 82–90.

37. Esat Soylu (engineer, Urteknik), interview with the author, 26 November 2010; Gürcan, interview; Tuna, interview; Tuncay, interview.

38. Alehar, "Türk Kahvesi Makinaları," *Kiva Han Forum*, 5 October 2010, http://www.kivahan.com.tr/forum/showthread.php?t=978 (accessed 4 January 2010). The user with the nickname Alehar introduced himself as the "inventor" of Dibek. My further research showed that the user had indeed been one of the engineers on the project. This was further confirmed by the way the user wrote about the project since his descriptions and concerns were consistent with the others I interviewed. I will be using the pseudonym Nihat Gözcü to refer to him in the article.

39. Soylu, interview.

40. Alehar, "Türk Kahvesi Makinaları."

41. Erhan Baygun (junior product designer, Urteknik), interview with the author, 17 October 2010; Tuna, interview; Soylu, interview.

42. Alehar, "Türk Kahvesi Makinaları."

43. Callon, "The Sociology of an Actor-Network," 29.

44. Akrich and Latour, "A Summary of a Convenient Vocabulary," 259.

45. Cansu, interview; Keskin, interview; Tuncay, interview. On projected, represented, and real users, see Johan Schot and Adri Albert de la Bruhèze, "The Mediated Design of Products, Consumption, and Consumers in the Twentieth Century," in *How Users Matter: The Co-construction of Users and Technology*, ed. Nelly Oudshoorn and Trevor Pinch (Cambridge, MA: MIT Press, 2003), 235; see also Steve Woolgar, "Configuring the User: the Case of Usability Trials," in *Sociology of Monsters? Essays on Power, Technology and Domination*, ed. John Law, pp. 57–102 (London: Routledge, 1991).

46. Baygun, interview.

47. Tuna, interview.

48. Tuna, interview; Alehar, "Türk Kahvesi Makinaları." Even though the question of gender falls outside of the scope of this article, it is necessary to state the gendered construction of this asymmetry as the engineer and his rational methods of abstraction and experimentation are placed as superior against the woman user and her domestic practices. Cynthia Cockburn and Susan Ormrod, *Gender and Technology in the Making* (London: Sage, 1993), 97.

49. Baygun, interview.

50. Tuna, interview.

51. Interview with Kunter Şekercioğlu, in Özgür Kayhan, "Yeni Ürün Geliştirme Sürecinde Tasarım İş Tanımı: Türkiye'deki Uygulamaların İrdelenmesi" (master's thesis, Istanbul Technical University, 2005).

52. Hugh Aldersley-Williams, *World Design: Nationalism and Globalism in Design* (New York: Rizzoli, 1992), 12, 14.

53. Viviana Narotzky, "Selling the Nation: Identity and Design in 1980s Catalonia," *Design Issues* 25, no. 3 (2009): 63.

54. Simon Jackson, "Sacred Objects: Australian Design and National Celebrations," *Journal of Design History* 19, no. 3 (2006): 252; Artemis Yagou, "Metamorphoses of Formalism: National Identity as a Recurrent Theme of Design in Greece," *Journal of Design History* 20, no. 2 (2007): 152.

Bibliography

Agnew, John. "The Territorial Trap: The Geographical Assumptions of International Relations Theory." *Review of International Political Economy* 1, no. 1 (1994): 53–80.

Akrich, Madeleine, and Bruno Latour. "A Summary of a Convenient Vocabulary for the Semiotics of Human and Nonhuman Assemblies." In *Shaping Technology/Building Society: Studies in Sociotechnical Change*, ed. Wiebe E. Bijker and John Law, pp. 259–264. Cambridge, MA: MIT Press, 1992.

Aldersley-Williams, Hugh. *World Design: Nationalism and Globalism in Design*. New York: Rizzoli, 1992.

Anderson, Benedict. *Imagined Communities*. Rev. ed. London: Verso, 2006.

Appadurai, Arjun. "Disjuncture and Difference in the Global Cultural Economy." In *Modernity at Large: Cultural Dimensions of Globalization*, pp. 27–47. Minneapolis: University of Minnesota Press, 1996.

———. "Playing with Modernity: The Decolonization of Indian Cricket." In *Modernity at Large: Cultural Dimensions of Globalization*, pp. 89–113. Minneapolis: University of Minnesota Press, 1996.

Appadurai, Arjun, Ashish Chadha, Ian Hodder, Trinity Jackman, and Chris Witmore. "The Globalization of Archaeology and Heritage: A Discussion with Arjun Appadurai." *Journal of Social Archaeology* 1, no. 1 (2001): 35–49.

Baudrillard, Jean. *Simulations*. New York: Semiotext(e), 1983.

Bazzal, Fadime Çoban. "En Başarılı 20 Yenilikçi Ürün." *Capital*, March 2007, 124–127.

Bell, David, and Gill Valentine. *Consuming Geographies: We Are Where We Eat*. London: Routledge, 1997.

Billig, Michael. *Banal Nationalism*. London: Sage, 1995.

Bora, Tanıl. *Medeniyet Kaybı: Milliyetçilik ve Faşizm Üzerine Yazılar*. Istanbul: Birikim Yayınları, 2006.

———. "Nationalist Discourses in Turkey." *South Atlantic Quarterly* 102, no. 2 (2003): 433–451.

Bowen, Sarah, and Marie Sarita Gaytán. "The Paradox of Protection: National Identity, Global Commodity Chains, and the Tequila Industry." *Social Problems* 59, no. 1 (2012): 70–93.

Brubaker, Rogers. "Rethinking Nationhood: Nation as Institutionalized Form, Practical Category, Contingent Event." *Contention* 4, no. 1 (1994): 3–14.

Bruner, Edward M. "Abraham Lincoln as Authentic Reproduction: A Critique of Postmodernism." *American Anthropologist* 96, no. 2 (1994): 397–415.

———. "The Maasai and the Lion King: Authenticity, Nationalism, and Globalization in African Tourism." *American Ethnologist* 28, no. 4 (2001): 881–908.

Callon, Michel. "The Sociology of an Actor-Network: The Case of the Electric Vehicle." In *Mapping the Dynamics of Science and Technology*, ed. Michel Callon, John Law, and Arie Rip, pp. 19–34. London: Macmillan Press, 1986.

———. "Some Elements of a Sociology of Translation: Domestication of the Scallops and the Fishermen of St Brieuc Bay." In *Power, Action and Belief: A New Sociology of Knowledge?*, ed. John Law, pp. 196–223. London: Routledge & Kegan Paul, 1986.

Charland, Maurice. "Technological Nationalism." *Canadian Journal of Political and Social Theory* 10, nos. 1–2 (1986): 196–220.

Cockburn, Cynthia, and Susan Ormrod. *Gender and Technology in the Making*. London: Sage, 1993.

Cohen, Eric. "Authenticity and Commoditization in Tourism." *Annals of Tourism Research* 15, no. 3 (1988): 371–386.

DeSoucey, Michaela. "Gastronationalism: Food Traditions and Authenticity Politics in the European Union." *American Sociological Review* 75, no. 3 (2010): 432–455.

Edgerton, David. *The Shock of the Old: Technology and Global History Since 1900*. London: Profile Books, 2008.

Featherstone, Mike. "Global Culture: An Introduction." In *Global Culture: Nationalism, Globalization and Modernity*, ed. Mike Featherstone, pp. 1–14. London: Sage, 1990.

———. "Islam Encountering Globalization: An Introduction." In *Islam Encountering Globalization*, ed. Ali Mohammadi, pp. 1–13. Oxon: RoutledgeCurzon, 2002.

Gellner, Ernest. *Nations and Nationalism*. Oxford: Blackwell, 1983.

Hannerz, Ulf. "Cosmopolitans and Locals in World Culture." In *Global Culture: Nationalism, Globalization and Modernity*, ed. Mike Featherstone, pp. 237–252. London: Sage, 1990.

Hughes, Thomas P. *Networks of Power: Electrification in Western Society, 1880–1930*. Baltimore: John Hopkins University Press, 1983.

Hamelink, Cees J. *Cultural Autonomy in Global Communication*. London: Longman, 1983.

Hugill, Peter J. *Global Communications Since 1844: Geopolitics and Technology*. Baltimore: Johns Hopkins University Press, 1999.

Jackson, Simon. "Sacred Objects: Australian Design and National Celebrations." *Journal of Design History* 19, no. 3 (2006): 249–255.

Kayhan, Özgür. "Yeni Ürün Geliştirme Sürecinde Tasarım İş Tanımı: Türkiye'deki Uygulamaların İrdelenmesi." Master's thesis, Istanbul Technical University, 2005.

Latour, Bruno. *Reassembling the Social: An Introduction to Actor-Network-Theory.* Oxford: Clarendon, 2005.

———. "Where Are the Missing Masses? The Sociology of a Few Mundane Artifacts." In *Shaping Technology/Building Society: Studies in Sociotechnical Change*, ed. Wiebe E. Bijker and John Law, pp. 225–258. Cambridge, MA: MIT Press, 1992.

Levitt, Theodore. *The Marketing Imagination.* London: Collier-Macmillan, 1983.

Lyth, Peter, and Helmuth Trischler. "Globalisation, History, and Technology: An Introduction." In *Wiring Prometheus: Globalisation, History, and Technology*, ed. Peter Lyth and Helmuth Trischler, pp. 7–20. Aarhus: Aarhus University Press, 2004.

MacCannell, Dean. *The Tourist: A New Theory of the Leisure Class.* Berkeley: University of California Press, 1999.

Männistö-Funk, Tiina. "The Crossroads of Technology and Tradition: Vernacular Bicycles in Rural Finland, 1880–1910." *Technology and Culture* 52, no. 4 (2011): 733–756.

Mazzarella, William. "'Very Bombay': Contending with the Global in an Indian Advertising Agency." *Cultural Anthropology* 18, no. 1 (2003): 33–71.

McGowan, Abigail S. "'All That Is Rare, Characteristic or Beautiful': Design and the Defense of Tradition in Colonial India, 1851–1903." *Journal of Material Culture* 10, no. 3 (2005): 263–287.

Millard, Andre. "Audio Cassette Culture and Globalisation." In *Wiring Prometheus: Globalisation, History, and Technology*, ed. Peter Lyth and Helmuth Trischler, pp. 235–250. Aarhus: Aarhus University Press, 2004.

Miller, Daniel. *Modernity: An Ethnographic Approach: Dualism and Mass-Consumption in Trinidad.* Oxford: Berg, 1994.

Morley, David, and Kevin Robins. *Spaces of Identity: Global Media, Electronic Landscapes and Cultural Boundaries.* London: Routledge, 1995.

Narotzky, Viviana. "Our Cars in Havana." In *Autopia*, ed. Peter Wollen and Joe Kerr, pp. 169–176. London: Reaktion Books, 2002.

———. "Selling the Nation: Identity and Design in 1980s Catalonia." *Design Issues* 25, no. 3 (2009): 62–75.

Ohmae, Kenichi. *Borderless World.* London: Collins, 1990.

Özkan, Derya, and Robert J. Foster. "Consumer Citizenship, Nationalism, and Neoliberal Globalization in Turkey: The Advertising Launch of Cola Turka." *Advertising and Society Review* 6, no. 3 (2005): doi:10.1353/asr.2006.0001.

"Plastik Kahve Pişiriciler Toplatılıyor." *Zaman*, 20 July 2008. http://www.zaman.com.tr/haber.do?haberno=716142 (accessed 10 December 2012).

Schiller, Herbert. *Communication and Cultural Domination.* New York: M. E. Sharpe, 1976.

Schot, Johan, and Adri Albert de la Bruhèze. "The Mediated Design of Products, Consumption, and Consumers in the Twentieth Century." In *How Users Matter: The Co-construction of Users and Technology*, ed. Nelly Oudshoorn and Trevor Pinch, pp. 229–246. Cambridge, MA: MIT Press, 2003.

Smil, Vaclav. *Prime Movers of Globalization: The History and Impact of Diesel Engines and Gas Turbines.* Cambridge, MA: MIT Press, 2010.

Smith, Anthony D. *Nationalism and Modernism: A Critical Survey of Recent Theories of Nations and Nationalism.* London: Routledge, 1998.

Terçin, Eda. "Ödüller Kalitemizi, Üretim Gücümüzü Ortaya Koyuyor." *Capital Online*, 1 March 2011. http://www.capital.com.tr/oduller-kalitemiziuretim-gucumuzu-ortaya-koyuyor-haberler/22797.aspx (accessed 8 January 2012).

Wimmer, Andreas, and Nina Glick Schiller. "Methodological Nationalism and Beyond: Nation-State Building, Migration and the Social Sciences." *Global Networks* 2, no. 4 (2002): 301–334.

Woolgar, Steve. "Configuring the User: The Case of Usability Trials." In *Sociology of Monsters? Essays on Power, Technology and Domination*, ed. John Law, pp. 57–102. London: Routledge, 1991.

Yagou, Artemis. "Facing the West: Greece in the Great Exhibition of 1851." *Design Issues* 19, no. 4 (2003): 82–90.

———. "Metamorphoses of Formalism: National Identity as a Recurrent Theme of Design in Greece." *Journal of Design History* 20, no. 2 (2007): 145–159.

CHAPTER 8

Canoes, Identity, and Globalization
The Story of Bill Mason's Camera Case

Bryan Dewalt

Director of Curatorial Division

Canada Science and Technology Museums Corporation Ottawa, Ontario, Canada

This chapter traces the biographies of a Canadian filmmaker, a Danish canoeist, and an object that passed between them. In 1986 Hans Michael Nielsen refashioned a small plastic shipping drum into a waterproof camera case and presented it to Bill Mason, the filmmaker, author, and artist whose books and films on canoeing skills and wilderness camping had done much to popularize these leisure activities in Canada and abroad. Mason's work is often considered to be quintessentially "Canadian." His canoeing practice and cinematic motifs were firmly rooted in the local and the historical, and most of his films were completed under the aegis of Canada's National Film Board (NFB), a federal agency charged with promoting national identity. And yet this gift from Scandinavia, which embodied modern industrial materials and had served time as a humble component in transnational supply chains, tied Mason into a global exchange of images and ideas.

The process of globalization, which in one sense has been gathering force since the sixteenth century, entered a new period of intensity in the decades after the World War II. On the one hand, this process assumed now familiar economic and political forms: for example, the spread of markets for goods, capital, and labor to all corners of the world and the rise of multinational corporations and nongovernmental organizations at the expense of nation states. At the same time, globalization was also a process of profound cultural change. "The idea of society as a self-contained, coherent, and clearly demarcated entity," largely coterminous with the nation-state, "was called into question."[1] Driven partly by movements in population but perhaps even more by the images and ideas carried by new communications technologies, the always precarious cultural integrity of the nation state was threatened. In this regard I am referring to Benedict Anderson's concept of the nation as an "imagined political community," an imaginative construction, aided by the rise of "print capitalism" and "national print languages" and abetted by states and protostates using these to build the foundations of their own legitimacy.[2] Canada was late to the nation-building game and hampered by linguistic duality, a multiethnic population, and an open border shared with the powerful United States of America. Nevertheless, in the twentieth century the Canadian state attempted to deploy the new technologies of broadcasting and film to articulate a shared national narrative. As the border-crossing carriers of images and ideas, however, these same technologies also brought into question the very idea of a shared narrative limited to the territory of a single state. According to the anthropologist Arjun Appadurai, "mediascapes" have become "repertoires" out of which anyone anywhere can form "scripts of imagined lives."[3] In the words of Jan Nederveen Pieterse, we have entered an era of "cultural hybridization," an "age of boundary crossing" when "the doors of erstwhile imagined communities open up."[4]

Scholars have for some time been interested in recycled objects, particularly as a form of creative expression practiced by third-world bricoleurs repurposing first-world detritus.[5] More recent critiques have taken this enthusiasm to task both for romanticizing the impoverished makers and for reading into their motivation an ironizing aesthetic that may exist more in the minds of Western art consumers than of the makers themselves.[6] Other observers point to the prevalence of recycling in all societies, not solely in the lands of the marginalized Other.[7] Nevertheless, the careers of such recycled objects as Bill Mason's camera case can still help shed light on globalization, particularly the transformations that things undergo as they move from the sphere of global commodities to that of singular objects.[8]

As several observers have suggested, globalization operates in multiple dimensions and scales. The dimensions can be material (the movement of people, capital, and goods) or immaterial (as in the flow of images, ideas, and cultural practices). In scale, it operates at a macrolevel (of nation-states, populations, corporations, and commodities) and a microlevel (of a single community, family, person, or object).[9] An object, moreover, can move both through space and through schemes of valuation, passing from the sphere of exchange to one where it has unique value outside the market. At the level of a single artifact, these physical and conceptual movements can be governed not just by large social forces and organizations but by small social groups and individuals.[10] Bill Mason's camera case, an object with well-documented travels and transformations, presents us with a unique opportunity to examine the multifaceted nature of globalization in more detail. This case study will explore Nielsen's gift as a globalized object, one that not only bobs on the currents of physical and virtual commerce but at each stage of its international career is actively transformed by its users and makers. In doing so I will examine how this object illuminates the relationship between nationalism and globalization in the late twentieth century.

An Unusual Case

The camera case itself is a fascinating study in hybridity, an obvious product of both industrial molding and skilled manual modification. The container (Figure 1) is almost cylindrical, about 30 cm in diameter at the base, tapering to slightly less at the top. With the lid secured, the entire case is 19 cm tall. The main body is a molded plastic shell, turquoise in color. It has two black plastic articulated ring handles on opposite sides, allowing it to be readily picked up or held in two hands. Molded in relief on the side is a pair of upward pointing arrows, of a type seen on the side of commercial packaging. The lid is also plastic, apparently of the same material as the sides and handles. Set into the top of the lid is a disk of blue, closed-cell foam that looks like the Ensolite material of sleeping mats used by campers. The underside of the lid, and in fact the entire interior of the case, is lined with buff-colored, open-cell foam (probably polyurethane). The lid fully lifts away from the case: it is not hinged, and there is no latch or buckle to close it. Instead, it is secured to the case by a galvanized steel ring that slips over raised edges on the lid and the mouth of the case. The ring is tightened by a spring clamp that, when set, ensures a seal between lid and case. Departing from the polymer materials that comprise the body and lid, the bottom of the case is made of varnished plywood secured to the body of the case by 20 steel screws that pass through the plastic sides into the wood.

The main body of the case is made of one of the most common man-made materials in the world: high-density polyethylene (HDPE). Developed in Britain before the World War II, HDPE

1.

A product of industrial molding and skilled manual modification. Photo by the author. Courtesy of Canada Science and Technology Museums Corporation.

is a thermoplastic that can be readily molded by heating and will retain its molded shape when cooled. Because of its general chemical resistance, toughness, flexibility, and ease of processing, it is used extensively in injection molding and blow molding of industrial drums, housewares, milk bottles and food containers, drug and toiletry packaging, and toys. It is even used to make canoes! Low-density polyethylene (LDPE), meanwhile, is found in bags, films, and bubble wrap, among many other products. By the early 1990s more polyethylene was being produced than any other plastic material in the world, and almost a third of this was being made outside the world's industrial heartlands of western Europe, North America, and Japan. It is truly a global material. The HDPE drum, meanwhile, has become a universal container that facilitates the global trade in food and chemical commodities.[11]

In its modified form, Bill Mason's particular HDPE drum appears admirably suited to its new purpose. Compact, watertight, and buoyant, it looks to have been built by someone who has spent a lot of time paddling canoes and portaging gear. It also bears several inscriptions that mark this former global commodity as a singular object, with its own unique biography. First, on the bottom of the case, burned into the center of the wood base is Hans Michael Nielsen's personal logo, the superimposed capital letters *M* and *H* surmounting the silhouette of a canoe (Figure 2):

2.
The varnished plywood bottom features Hans Michael Nielsen's personal logo. Photo by the author. Courtesy of Canada Science and Technology Museums Corporation.

a touch that is ironically reminiscent of the ubiquitous corporate identifiers on mass-produced commodities. Second, on the underside of the lid and on the outside of the case body, the owner has printed his name and telephone number in black permanent marker: BILL MASON 819-827 2282. Where this lettering appears on the exterior, it has faded or been worn away in places. In other signs of extensive use, the plastic sides are scuffed, and the finish on the wood bottom has worn away around the edges. Similarly, the screw heads show signs of rust, and there is a good deal of powdery white oxidation and flash rust on the galvanized steel ring. Inside, the open-cell foam has likely yellowed from its original state, and there are pieces of gray duct tape holding it in place, which may or may not be original. Finally, on the container, on the lid, and on the steel fastening ring are perhaps the ultimate markers of singularity in the first world, museum catalog numbers inked by hand.

A Life with Canoes and Cameras

William Clifford Mason was born in Winnipeg in 1929. Trained at the University of Manitoba's School of Art, he worked for several Winnipeg advertising agencies while acquiring skills in animation, filmmaking, and photography. He also worked summers as a canoe guide at a summer camp and undertook the first of many backcountry trips where he taught himself the rudiments of white water canoeing and wilderness camping.[12] In 1958 Mason left Winnipeg to take a job as an animator with Crawley Films in Ottawa. Initially living in a tent at Meech Lake, Mason worked on a television cartoon series while moonlighting as art director for a Winnipeg advertising agency. His breakthrough as a filmmaker came with *Paddle to the Sea* (1966), a film based on the 1941 children's book by American author Holling C. Holling. Produced as a freelance project for the NFB, *Paddle* was nominated for an Academy Award in 1968. It was followed by a string of acclaimed documentaries also made for the NFB. In the 1970s Mason made a series of instructional films, *Path of the Paddle* (1977), that have become classic texts on the art of canoeing. These were followed by a number of other productions culminating in *Waterwalker* (1984), in which Mason integrated his love of the wilderness and the canoe with his deeply held spiritual beliefs (Figure 3). After completion of *Waterwalker* Mason retired from filmmaking to paint. He was diagnosed with terminal cancer in 1989 and died a few months later, aged just 59. Over the course of his career, Mason had an impact beyond Canada's borders. He received close to 50 international awards for his films, and his book based on the instructional films, *Path of the Paddle*, was translated into French and German.

Constructing Canadian Identity

Mason's films resonated with a number of social and cultural trends in the middle decades of the twentieth century. Beginning in the late nineteenth century, naturalists, artists, educators, and religious thinkers promoted wilderness tourism and outdoor education as an antidote to the spiritual ennui and physical debility caused, they believed, by sedentary urban life. As early as the turn of the twentieth century, canoes (and faux aboriginal names, campfire tales, and regalia)

3.
Bill Mason paddling on Old Woman Bay, Lake Superior, 1980s. Photo by Paul Mason. Courtesy of the Mason Family.

figured in the summer camp movement across North America. This was part of a decades-long international trend, at least in the Anglo-Saxon world, that spawned the YMCA, the Boy Scouts, and the national parks movement. Among the notable Canadian exponents of this turn to the wilderness were the writers Ernest Thompson Seton (author of *Two Little Savages*, among others) and Grey Owl (the pseudonym of English expatriate and Indian impostor, Archie Belaney) and the landscape painters Tom Thomson and the Group of Seven.[13]

In his youth, Mason was introduced to this spirit through his art training, his religious education, and his summer experiences as camper and guide at the Intervarsity Christian Fellowship's Manitoba Pioneer Camp. He was also greatly influenced by the writing of Calvin Rutstrum, particularly his backcountry manual, *The Way of the Wilderness* (1946).[14] Mason's first solo production, *Wilderness Treasure*, began as a promotional film for the Pioneer Camps but evolved into a cinematic narration of a canoe trip suffused with reverence for the natural world. In the 1960s and 1970s, the wilderness tradition was reinvigorated by the emergence of the global environmental movement. In films like *Paddle to the Sea* and *The Rise and Fall of the Great Lakes* Mason captured nascent public concern for pollution and the destruction of wildlife habitat.

Mason's canoeing films emerged at the confluence of this wilderness tradition and global environmental consciousness with a third major movement of the mid-twentieth century, English-Canadian nationalism. This was not self-consciously expressed by Mason, whose interests beyond canoes, paddles, and waterways were more spiritual than patriotic. Nevertheless, many Canadian nationalists looking for a firm foundation on which to construct a secure Canadian identity in an increasingly fluid global cultural landscape took inspiration from the fur trade past and from the land, particularly the boreal forests and abundant waters of the rugged Precambrian Shield.[15] These provided a pool of historical and geographical signifiers with which to express an indigenous national essence. The canoe and the fur trade were deeply embedded in a midcentury academic historiography that informed official and popular histories. In *The Fur Trade in Canada* (1930) Harold Innis argued "Canada emerged as a political entity with boundaries largely determined by the fur trade. . . . The present Dominion emerged not in spite of geography but because of it."[16] In Donald Creighton's *Commercial Empire of the St. Lawrence* (1937), the dual geographic features of Precambrian Shield and St. Lawrence River figured as "the bone and the blood-tide of the northern economy."[17] The east–west flow of commerce along the waterways through the shield became the foundation for the Canadian state. Central to this nascent Canadian state was an alliance of Europeans and indigenous peoples that in the hinterland regions formed a "blent society where Europe and America met and mingled."[18] This potent founding narrative for the Canadian state, one that emerged out of the rock and rivers of the land itself, linked by kinship and trade to the land's first peoples, was shared in the popular literature of canoeing and made its way into the official discourse on nationhood.[19]

In the twentieth century Canada's ties to the globe-spanning British Empire weakened as the empire itself withered. In its place, already existing economic, cultural, and political links to the United States grew stronger and, with them, a fear of economic dependence, political subservience, and cultural assimilation. Conversely, Canada's contribution to the Allies' victories in two world wars stimulated national pride while promoting a sense of membership in a larger community of nations. At the same time, Canada became a magnet for millions of non-British immigrants who both aided the growth of Canada's trade-dependent capitalist economy and brought into question the old cultural and linguistic accommodations between French and British settler communities. These accommodations were equally threatened by an increasingly assertive nationalism among the French-speaking population of the province of Quebec, who saw their destiny not in a united Canada but in an independent country of their own.[20]

Into this context of assertiveness and vulnerability, nationalism and globalization, came the canoe. Long after they had ceased to figure much in the actual economy of the country, canoes and canoe tripping began to acquire a new status in the official construction of Canadian identity. Canoes and canoeists began to appear on postage stamps and coins,[21] and the federal government began to encourage canoe tripping as a leisure activity.[22] By the 1960s the latter had gone beyond tourist promotion to the official deployment of canoeing as a form of historical reenactment. In 1965 Bill Mason attended the initial organizing meeting for a government-sponsored cross-Canada canoe "pageant," in which, two years later, teams marked the centennial of the Canadian confederation by racing across the country in modern canvas-covered craft painted to resemble the birch bark *canots du Nord* of the fur trade era.[23] In 1969 the National and Historic Parks

Branch issued the first volume in a series on major historic themes, *Fur Trade Canoe Routes of Canada: Then and Now*. Although largely a descriptive account of the history and geography of major voyageur routes that was aimed at adventurous canoeists, the author Eric Morse drew attention in his conclusion to a nation-building moral: "In a decade when some Canadians are vying with one another to discover and exploit their differences, it is healthful to review a story in which French and English, Indians, Métis, and Scots all worked closely together. . . . The fur trade . . . was a vast Canadian enterprise."[24]

The National Film Board, charged with "interpreting Canada to Canadians and other nations," was actively involved in this business of identity construction through its extensive production of documentaries, educational films, and television shows.[25] The first NFB production on which Mason worked was *The Voyageurs*. His *Path of the Paddle* series for the NFB may be read as an instruction manual for the citizen as canoeist. The films of Bill Mason and the activities and practices he promoted became integral parts of national life. Screened in schools, on television, in theatres, and in other venues, Mason's films have been watched by millions of Canadians from the 1960s through to the present day. They continue to be available on DVD or by download or streaming from the National Film Board website.

In Canada, the canoe was transformed, in the words of Misao Dean, from "a simple means to get from one place to another" into "an object, and a pastime, that is assumed to have meaning and significance in itself."[26] The canoe became a kind of "ideology"[27] that simultaneously appropriated and effaced aboriginal origins. Like the wagon for Afrikaner nationalists, to borrow Anne McClintock's formulation, the canoe became a "fetish object," deployed in "commodity spectacles" like the Centennial Canoe Pageant, where the act of canoeing became a reenactment that affirmed the integrity of the Canadian state.[28] In the twenty-first century, the canoe, according to Bruce Erickson, continues to play a key role in narrating national identity, figuring both as "a boat that has arisen out of nature and as a boat that connects the nation with nature."[29] The canoe has become a "technology of identity" that "naturalizes" a particular vision of Canada (for example, of a tolerant, multicultural, and capitalist society) and effectively obscures histories of contention, domination, and exploitation. The process of naturalization, moreover, is all the more powerful for the way that recreational canoeing allows Canadians to identify with the dominant vision while engaging in a routine, seemingly apolitical leisure activity.[30] Although the canoe figures in private pastimes and personal narratives, at the same time it bears the weight of a certain kind of "Canadianness." It is to recreational canoeists, real and aspiring, that the films and books of Bill Mason are particularly compelling.

From Global Supply Chain
to Canadian Canoe Route

Considering the elaborate effort involved in the appropriation of the canoe as a national symbol, Mason received his camera case from an unlikely source. Hans Michael Nielsen was born in a suburb of Copenhagen in 1955. As a boy, he relates, he became "very interested in the American Indians and their canoes as well."[31] He first sat in a canoe at the age of 15. At 18, around the time he began an apprenticeship as a carpenter, he took his first canoe trip in Sweden. A year later he bought a small cabin in Sweden and purchased a fiberglass canoe that he assembled himself. Nielsen was employed as a carpenter from 1976 to 1994, after which he trained as a teacher. He now works as a carpenter and teaches outdoor education at the University of Copenhagen. Nielsen has constructed at least two other canoes, as well as paddles and various pieces of camping gear. In 1987 he led a canoe trip in northern Sweden on the Tornio River between Finland and Sweden in a 9-meter "voyageur" canoe of his own construction. This journey was captured in a film he made, *En stor kano på en fri elv* (A Large Canoe on a Free River).[32]

Although he later watched Bill Mason's entire canoe film oeuvre, Nielsen first encountered Mason through his book, *Path of the Paddle*, which he purchased in a bookstore in Alaska after a canoe trip there in 1982. In his words, it became his "canoeing bible" and led him to completely alter his paddling technique. "The book made the first ripples that would shape the way we paddle in Scandinavia today."[33] It also inspired him to travel to Canada's Northwest Territories in 1986 to paddle the Nahanni River, Canada's first UNESCO World Heritage Site. While on the Nahanni, Nielsen paddled with Mason's son, Paul, and later met the father at the national park office in Fort Simpson. It was at this time that Mason noticed and expressed admiration for a distinctive camera case that Nielsen had made. On a visit to Mason's home at Meech Lake, Nielsen presented Mason with two handmade paddles. Once back in Copenhagen, he made a case for Mason, which he then sent as a gift.[34]

In the 1970s canoeists had begun to use HDPE containers for dry storage of food and other items on backcountry canoe trips. Light, tough, and waterproof, these first "canoe barrels" were, in fact, reused drums and pails once employed for shipping olives, paints, pharmaceuticals, and the like. According to Wally Schaber, an Ottawa-area canoe outfitter who was a friend of Mason, canoeists on both sides of the Atlantic hit upon the idea by the mid-1980s of fitting barrels with backpack harnesses for ease of portaging. Schaber purchased his first lot of new, rather than repurposed, drums after attending a trade show in Britain. Today, purpose-built canoe barrels (Figure 4) are readily available for purchase from outdoor equipment suppliers.[35]

In building the camera case, Nielsen took the process of adaptation one step further. He cut the top third off a used 30-L drum that he had obtained from a factory near his Copenhagen home and constructed a new floor from 18-mm birch plywood that he shaped on a lathe and screwed to the bottom edges of the truncated drum. To ensure water tightness he sealed the interface between the two pieces with a rubber O-ring and silicone caulking and also caulked the screw holes. The height of the finished case allowed it to fit under the seat of a canoe, and with the Ensolite pad on the lid it could be used as a stool in camp. By threading a webbing strap through the handles it could be carried over the shoulder or tied to a canoe thwart. In 1987 on his journey down the Tornio River, Nielsen lost his own case when his canoe capsized. Thinking it gone for good, he was delighted to find it floating in an eddy 18 km downstream. Despite having tumbled

4.
Canoe barrels with and without harness, Trailhead Paddle Shack, Ottawa, 2014. Photo by the author. Courtesy of Canada Science and Technology Museums Corporation.

through several rapids on its solitary journey, the interior of the case was still dry. Nielsen made a number of cases for himself and friends, and in the early 1990s he even considered commercial production.[36]

Coming of age in the "pre-GORE-TEX" era, Mason relied extensively on traditional materials like cedar, canvas, leather, and wool. These remained central to his film aesthetic in the 1970s and 1980s, by which time other outdoor enthusiasts had largely embraced the various polymers and synthetic fabrics so common in the backcountry today. Mason was also an inveterate bricoleur, routinely repairing and modifying his tools with duct tape, cloth hockey tape, and the shafts of wooden hockey sticks. For filming whitewater canoe scenes he made extensive use of a waterproof camera case fashioned from a war surplus steel ammunition box. Despite the space-age materials, therefore, Nielsen's industrial drum-canoe barrel-camera case was a fitting tribute to Mason's practice. According to his friend and long-time camera operator, Ken Buck, Mason used the case in the final two years of his life to carry one or two Nikon 35mm SLR cameras and his Nikkor lenses.[37] The Canadian filmmaker was very pleased with the Danish container's practicality, in one letter telling Nielsen "it is easy and fast to open and I can have my camera with a 300mm lens on ready for shooting."[38] In another letter he reported, "I have been getting into

5.
The case, complete with SLR camera, appeared in Mason's wilderness camping manual, *Song of the Paddle* (1988). It is labelled 'I' in the photo. Immediately to its right: a repurposed olive barrel. Courtesy of the Mason Family.

some huge stuff on the Ottawa [River] and can now roll the canoe upright after an upset. Great fun and your camera case is absolutely water tight and indestructible. I sure gave it a test."[39] Mason later featured the case in the text and illustrations of his 1988 wilderness camping manual, *Song of the Paddle* (Figure 5), calling it "as clever as anything I've ever seen."[40]

Canoes Crossing Boundaries

Judging from the source of the gift, the shared passion for canoeing, and even the equitable nature of the exchange in expertise (Mason's technique for Nielsen's tools), it is clear that canoeing and the potentially divergent meanings that are assigned to it cannot be contained within the fluctuating historical borders of the Canadian state. International air travel has enabled Scandinavian canoeists to consider the Nahanni River their own backyard. Even within North America, the iconic birch bark canoe, and the network of small waterways to which it was adapted, had historically existed on both sides of the line that ultimately defined the Canada–United States border. Calvin Rutstrum, the young Bill Mason's idol, was the American son of Swedish immigrants who spent many years paddling the Boundary Waters region of Minnesota and Ontario and whose book *The Way of the Wilderness* was a classic text of the *American* wilderness tradition running from Henry David Thoreau through Ansel Adams.[41] The first comprehensive material culture study of canoes, *The Bark Canoes and Skin Boats of North America*, was coauthored by Edwin Tappan Adney, an Ohio-born, naturalized Canadian, and Howard I. Chapelle, curator of transportation at the American Museum of History and Technology.[42]

In the late nineteenth century, many affluent Americans became avid canoeists. Organized in 1879, the American Canoe Association (ACA) held annual meets in waters near, astride, or across the Canadian border. For many years, in fact, Canadians were members of the Northern Division of the ACA.[43] The modern canoe-building industry, meanwhile, owed a great deal to the innovations of American manufacturers, most notably the canvas-covered craft developed in Maine in the 1880s and the aluminum shells introduced by the American aircraft maker Grumman after World War II. Mason's cherished Chestnut Prospector cedar-canvas canoe, for example, was a Canadian copy of an American design.[44]

Even assuming this broader, cross-border homeland, canoes did not stay put. Probably the single most important vector for the beginning of this modern global journey was a Scotsman, John MacGregor, who had been intrigued by the indigenous boats of Canada and Siberia during trips to those regions in the 1850s. In 1865 he ignited an international canoeing craze by making a solo journey through the waterways of Belgium, Germany, Switzerland, and France in his odd, oak-planked, cedar-decked canoe, the *Rob Roy*, which fused North American forms with English boat-building technology. MacGregor was greeted by curious crowds wherever he went, and the record of this journey, *A Thousand Miles in the Rob Roy Canoe* (1866), was an international bestseller (Figure 6).[45] In the following year MacGregor toured Norway, Sweden, and Denmark, a trip recounted in his book *The Rob Roy on the Baltic* (1867). Over the ensuing decades, the primary European expression of canoeing was athletic. In 1866 the Royal Canoe Club was formed with MacGregor as its captain, and canoe racing became a popular European sport. In 1924 the forerunner of the International Canoe Federation (ICF) was established in Copenhagen.[46] The open-decked canoe propelled with a single-bladed paddle has for decades been known for competitive purposes as the "Canadian canoe." Yet throughout its history as an Olympic sport, which began in Berlin in 1936, European canoeists have consistently outraced their Canadian competitors, accounting for more than 90% of all medals awarded.[47]

MacGregor's colorful exploits notwithstanding, it is likely that many Danes would already have known something about the indigenous boats of Greenland, their New World colony. In 1867, MacGregor remarked that his Danish hosts preferred the Greenland word *kayak* to describe the decked craft that he propelled with a twin-bladed paddle, and the Danes exceeded all others for their "obstinate inquisitiveness" about his boat.[48] Denmark was, after all, an imperial power deeply implicated in a network of interactions that embraced all of northern Europe and North America. The modern European canoe and kayak represent the transfer of native North American technologies to the colonial centers, where they have become indigenized and incorporated into hybrid cultural practices like racing.

6.
John MacGregor portaging the *Rob Roy*, southern Germany, 1865. From *A Thousand Miles in the Rob Roy Canoe* (1866). Image reproduced from the Project Gutenberg ebook.

Beyond the canoe itself, twentieth-century global mediascapes built around publishing, film, and television technologies offered a full repertoire of images from which their consumers could script alternate lives and fashion new social bonds. Fed by popular images of indigenous North Americans, for example, "Indian Hobbyist" clubs formed in several European countries, offering enthusiasts the opportunity to gather at "powwows" to dress, dance, and sing like "red Indians." Largely because of the popular stories of Karl May (1842–1912), Indian hobbyism has been particularly popular in Germany. An *Indianklubben* was also formed in Sweden in 1959, and one of its longstanding members, Erik Englund, wrote a book for Danish schools that was published in 1971.[49] It is possible that this was one source of Hans Michael Nielsen's "interest in the American Indians and their canoes as well." This European fascination with the idealized lives of North American natives may also have intermingled with powerful local traditions celebrating life lived close to nature. As Simon Schama has made clear, there is nothing uniquely Canadian about a national origin myth rooted in landscapes and waterways.[50] The forest and the noble savage that purportedly inhabited it were at the core of German nationalism since the Renaissance and more generally fed the imagination of romantics everywhere from the eighteenth century onward.[51]

Shorn of national mythology, the canoe, and images of it, can feed an abiding passion that knows no borders. To Hans Michael Nielsen, the appeal is both simple and personal. "The canoe is a fantastic craft, which enables you to go to remote places and get close to nature." Yet, unintentionally echoing Misao Dean, he writes, "To me the canoe is more than just a means of transportation. . . . It has become a big part of my life."[52] In an age of global flows of commodities and images, this life finds value in unique, handmade objects and the companionship of fellow wilderness travelers. It is possible to see Bill Mason's camera case as a kind of token that solidified the bond between new world and old. By giving it to Mason, Nielsen asserted not only friendship or respect for a master paddler but also their common membership in a canoeing community that transcended national borders. There is also an intriguing symmetry in the gift itself. The colorful industrial drum molded from ubiquitous polyethylene can be found on loading docks and shop floors around the world. It is the antithesis of the birch bark canoe. Yet this global object has been reimagined and rebuilt for very personal reasons and for use in very specific local circumstances. It is a hybrid, both mass produced and handmade, local and global. One thing it is not, however, is national. As an object it is essentially stateless, and this underlines the fragility of the Canadian nation-building project, for which the frail, versatile canoe carried so much freight in a century swept by the currents of globalization.

Conclusion

Whatever his own personal motivations, Bill Mason's work became part of a larger nation-building project in which the canoe figured as an icon in a narrative of national origins. Yet the example of Bill Mason's camera case suggests that this constructed sense of national belonging is not watertight. Symbolic objects and practices that we associate with the nation can be appropriated by others not part of the nation and can be used to construct other "imagined communities" that cut across and potentially weaken national boundaries. Canoes, by their very nature, are mobile technologies, and it is somehow fitting that they figure in a story of globalization. Canoes are also containers that have historically carried all manner of cargo, from furs to spiritual yearnings to national aspirations. In this sense there is a fascinating symmetry between the canoe and Bill Mason's camera case. The substance of the case is plastic; it can be shaped by human agency into multiple possible forms. As an industrial drum it has contained and carried a variety of possible substances across great distances. In its final incarnation it carried Japanese cameras that themselves produced and contained images or, rather, latent images that were further processed and, in their consumption, transformed again. As a gift, it also carried something else, an expression of friendship and assertion of community that was at once intimate and global. Ultimately, as the node at which multiple streams converge, Bill Mason's camera case is a unique expression of the dimensions and scales of globalization.

Notes

1. Jürgen Osterhammel and Niels P. Petersson, *Globalization: A Short History* (Princeton, NJ: Princeton University Press, 2005), 14. A further overview of the history and historiography of globalization may be found in Bruce Mazlish, *The New Global History* (New York: Routledge, 2006).

2. Benedict Anderson, *Imagined Communities: Reflections on the Origin and Spread of Nationalism*, rev. ed. (London: Verso, 2006), 6, 42–43.

3. Arjun Appadurai, "Disjuncture and Difference in the Global Cultural Economy," *Public Culture* 2, no. 2 (Spring 1990): 9.

4. Jan Nederveen Pieterse, *Globalization and Culture: Global Mélange*, 2nd ed. (New York: Rowman & Littlefield, 2009), 88.

5. See, for example, Allen F. Roberts, "Chance Encounters, Ironic Collage," *African Arts* 25, no. 2 (April 1992): 55–56.

6. Corinne A. Kratz, "Rethinking Recyclia," *African Arts* 28, no. 3 (Summer 1995): 7–8.

7. Jeremy Coote, Chris Morton, and Julia Nicholson, *Transformations: The Art of Recycling* (Oxford: Pitt Rivers Museum, 2000), 8.

8. For more on the concepts of commodity and singular objects, see Igor Kopytoff, "The Cultural Biography of Things: Commoditization as Process," in *The Social Life of Things: Commodities in Cultural Perspective*, ed. Arjun Appadurai (Cambridge: Cambridge University Press, 1988), 64–91.

9. Appadurai, "Disjuncture and Difference," 6–7, 17; Osterhammel and Petersson, *Globalization*, 21–22.

10. Kopytoff, "Cultural Biography of Things," 79–81.

11. G. D. Wilson, "Polythene: The Early Years," in *The Development of Plastics*, ed. Susan Mossman and Peter Turnbull (London: Royal Society of Chemistry, 1994), 70–85; J. A. Brydson, *Plastics Materials*, 6th ed. (Oxford: Butterworth Heinemann, 1995), 201–203, 218–219, 237–238.

12. For the authoritative biography of Mason, see James Raffan, *Fire in the Bones: Bill Mason and the Canadian Canoeing Tradition* (Toronto: Harper Collins, 1996).

13. Raffan, *Fire in the Bones*, 14–15, 193–198; Daniel Francis, *National Dreams: Myth, Memory, and Canadian History* (Vancouver: Arsenal Pulp Press, 1997), 132–149.

14. Raffan, *Fire in the Bones*, 55–58.

15. This massive, glacier-scoured geological formation covering almost one-half of Canada's land area is more commonly known in the country as the Canadian Shield.

16. Harold Innis, *The Fur Trade in Canada: An Introduction to Canadian Economic History*, rev. ed. (Toronto: University of Toronto Press, 1956), 393.

17. Donald Creighton, *The Empire of the St. Lawrence* (Toronto: Macmillan, 1956), 5.

18. Creighton, *Empire of the St. Lawrence*, 16.

19. The dust jacket to Ronald H. Perry's 1948 canoe manual for summer campers stated that "the very existence of our national life is to a great extent attributable to the canoeing exploits of our ancestors during the early days of exploration and the fur trade." Ronald H. Perry, *The Canoe and You* (Toronto: Dent, 1948).

20. For a general introduction to this period in Canadian history, see Desmond Morton, *A Short History of Canada*, 5th ed. (Toronto: McClelland & Stewart, 2001).

21. James Raffan, *Bark, Skin and Cedar: Exploring the Canoe in Canadian Experience* (Toronto: Harper Collins, 1999), 241; Royal Canadian Mint, "Limited Edition Proof Silver Dollar—75th Anniversary of the First Canadian Silver Dollar (2010)," http://www.mint.ca/store/product/product.jsp?print=true&itemId=prod930001 (accessed 8 March 2013).

22. See, for example, Canada, Ministry of the Interior, *Voyages en Canot dans le Québec* (Ottawa: Ministry of the Interior, 1931).

23. Misao Dean, "The Centennial Voyageur Canoe Pageant as Historical Re-enactment," *Journal of Canadian Studies/Revue d'Études canadienne* 40, no. 3 (Fall 2006): 43–67.

24. Eric W. Morse, *Fur Trade Canoe Routes of Canada: Then and Now* (Ottawa: Queen's Printer, 1969), 117. Significantly, the foreword to this book was written by Pierre Elliott Trudeau, the newly elected prime minister of Canada. Trudeau's French-Scottish parentage embodied Canada's linguistic duality, and his political program emphasized "national unity" in the face of emerging separatist threats in Quebec. Trudeau was also an avid canoeist, having learned to paddle as a boy at summer camp, and was frequently filmed and photographed paddling his canoe on a northern lake. His lakeside summer residence in the Gatineau Hills was just down the road from Mason's home, and the two became good friends. See Raffan, *Fire in the Bones*, 168; Francis, *National Dreams*, 143–144.

25. Rodney C. James, *Film as a National Art: NFB of Canada and the Film Board Idea* (New York: Arno Press, 1977), 709; see also Gary Evans, *In the National Interest: A Chronicle of the National Film Board of Canada from 1949 to 1989* (Toronto: University of Toronto Press, 1991).

26. Dean, "Centennial Voyageur Canoe Pageant," 49.

27. Francis, *National Dreams*, 128.

28. Anne McClintock, *Imperial Leather: Race, Gender and Sexuality in the Colonial Contest* (New York: Routledge, 1995), 370–376.

29. Bruce Erickson, *Canoe Nation: Nature, Race, and the Making of a Canadian Icon* (Vancouver: University of British Columbia Press, 2013), xiii.

30. Erickson, *Canoe Nation*, 12–15. Heavily influenced by Michel Foucault and Homi Bhabha, Erickson's work is the most exhaustive exploration to date of the iconography of the canoe.

31. Hans Michael Nielsen, e-mail message to author, 30 June 2012.

32. Kanokram, "I solokano på grænseelven," http://www.heltude.dk/kanokram.htm (accessed 8 March 2013).

33. Nielsen, e-mail message to author, 30 June 2012.

34. Nielsen, e-mail message to author, 30 June 2012.

35. Wally Schaber, e-mail message to author, 27 May 2012. For examples of canoe barrels currently sold through retail channels, see Trailhead Paddle Shack website, http://trailheadpaddleshack.ca/store/paddling-dry-storage (accessed 29 January 2016).

36. Nielsen, e-mail message to author, 30 June 2012.

37. Ken Buck, interview by the author, 16 November 2009.

38. Nielsen, e-mail message to author, 24 May 2010.

39. Bill Mason to Hans Michael Nielsen, n.d., original in possession of Mrs. Joyce Mason.

40. Bill Mason, *Song of the Paddle: An Illustrated Guide to Wilderness Camping* (Toronto: Key Porter, 1988), 102–103.

41. Wikipedia, "Calvin Rutstrum," https://en.wikipedia.org/wiki/Calvin_Rutstrum (accessed 3 May 2012).

42. Raffan, *Bark, Skin and Cedar*, 66–67; Edwin Tappan Adney and Howard I. Chapelle, *The Bark Canoes and Skin Boats of North America* (Washington, D.C.: Smithsonian Institution, 1964).

43. Lee J. Vance, "Canoeing in America," *Cosmopolitan Magazine*, October 1893, 711. At the annual meets Canadians were noted for their prowess as paddlers, whereas their southern neighbors excelled at canoe sailing.

44. Raffan, *Bark, Skin and Cedar*, 67–70, 155; Raffan, *Fire in the Bones*, 196.

45. John MacGregor, *A Thousand Miles in the Rob Roy Canoe* (1866; repr., Murray, UT: Dixon-Price, 2000), 5; Raffan, *Fire in the Bones*, 62.

46. International Canoe Federation, "History," http://www.canoeicf.com/history (accessed 29 January 2016).

47. In recent years, Canadians have been rivaled also by paddlers from China, Australia, and New Zealand. Wikipedia, "List of Olympic Medalists in Canoeing (Men)," https://en.wikipedia.org/wiki/List_of_Olympic_medalists_in_canoeing_(men) (accessed 25 January 2013).

48. John MacGregor, *The Rob Roy on the Baltic*, 2nd ed. (Boston: Roberts Bros., 1867), 193–194.

49. Colin F. Taylor, "The Indian Hobbyist Movement in Europe," *Handbook of North American Indians,* ed. William C. Sturtevant, vol. 4: *History of Indian-White Relations*, ed. Wilcomb E. Washburn (Washington, D.C.: Smithsonian Institution Press, 1988), 567.

50. By naming his various canoes the *Rob Roy*, for example, John MacGregor linked his pastime to the Jacobite folk hero of the Scottish highlands, sometimes referred to as the Scottish Robin Hood.

51. Simon Schama, *Landscape and Memory* (Toronto: Random House, 1995), 15, 91–103.

52. Nielsen, e-mail message to author, 30 June 2012.

Bibliography

Adney, Edwin Tappan, and Howard I. Chapelle. *The Bark Canoes and Skin Boats of North America.* Washington, D.C.: Smithsonian, 1964.

Anderson, Benedict. *Imagined Communities: Reflections on the Origin and Spread of Nationalism.* Rev. ed. London: Verso, 2006.

Appadurai, Arjun. "Disjuncture and Difference in the Global Cultural Economy." *Public Culture* 2, no. 2 (Spring 1990): 1–24.

Brydson, J. A. *Plastics Materials.* 6th ed. Oxford: Butterworth Heinemann, 1995.

Canada. Ministry of the Interior. *Voyages en Canot dans le Québec.* Ottawa: Ministry of the Interior, 1931.

Coote, Jeremy, Chris Morton, and Julia Nicholson. *Transformations: The Art of Recycling.* Oxford: Pitt Rivers Museum, 2000.

Creighton, Donald. *The Empire of the St. Lawrence.* Toronto: Macmillan, 1956. Originally published as *The Commercial Empire of the St. Lawrence, 1760–1850.* Toronto: Ryerson Press, 1937.

Dean, Misao. "The Centennial Voyageur Canoe Pageant as Historical Re-enactment." *Journal of Canadian Studies/Revue d'Études canadienne* 40, no. 3 (Fall 2006): 43–67.

Erickson, Bruce. *Canoe Nation: Nature, Race, and the Making of a Canadian Icon.* Vancouver: University of British Columbia Press, 2013.

Evans, Gary. *In the National Interest: A Chronicle of the National Film Board of Canada from 1949 to 1989.* Toronto: University of Toronto Press, 1991.

Francis, Daniel. *National Dreams: Myth, Memory, and Canadian History.* Vancouver: Arsenal Pulp Press, 1997.

Innis, Harold A. *The Fur Trade in Canada: An Introduction to Canadian Economic History.* 1930. Rev. ed. Toronto: University of Toronto Press, 1956.

James, Rodney C. *Film as a National Art: NFB of Canada and the Film Board Idea.* New York: Arno Press, 1977.

Jennings, John, Bruce W. Hodgins, and Doreen Small. *The Canoe in Canadian Cultures.* Toronto: National Heritage, 1999.

Kopytoff, Igor. "The Cultural Biography of Things: Commoditization as Process." In *The Social Life of Things: Commodities in Cultural Perspective*, ed. Arjun Appadurai, pp. 64–91. Cambridge: Cambridge University Press, 1988.

Kratz, Corinne A. "Rethinking Recyclia." *African Arts* 28, no. 3 (Summer 1995): 1, 7–8, 10–12.

MacGregor, John. *The Rob Roy on the Baltic.* 2nd ed. Boston: Roberts Bros., 1867.

———. *A Thousand Miles in the Rob Roy Canoe.* 1866. Reprint, Murray, UT: Dixon-Price, 2000.

Mason, Bill. *Song of the Paddle: An Illustrated Guide to Wilderness Camping.* Toronto: Key Porter, 1988.

McClintock, Anne. *Imperial Leather: Race, Gender and Sexuality in the Colonial Contest.* New York: Routledge, 1995.

Mazlish, Bruce. *The New Global History.* New York: Routledge, 2006.

Morse, Eric W. *Fur Trade Canoe Routes of Canada: Then and Now.* Ottawa: Queen's Printer, 1969.

Morton, Desmond. *A Short History of Canada.* 5th ed. Toronto: McClelland & Stewart, 2001.

Osterhammel, Jürgen, and Niels P. Petersson. *Globalization: A Short History.* Princeton, NJ: Princeton University Press, 2005.

Perry, Ronald. *The Canoe and You.* Toronto: Dent, 1948.

Pieterse, Jan Nederveen. *Globalization and Culture: Global Mélange.* 2nd ed. New York: Rowman & Littlefield, 2009.

Raffan, James. *Bark, Skin and Cedar: Exploring the Canoe in Canadian Experience.* Toronto: Harper Collins, 1999.

———. *Fire in the Bones: Bill Mason and the Canadian Canoeing Tradition.* Toronto: Harper Collins, 1996.

Roberts, Allen F. "Chance Encounters, Ironic Collage." *African Arts* 25, no. 2 (April 1992): 54–63, 97–98.

Schama, Simon. *Landscape and Memory.* Toronto: Random House, 1995.

Taylor, Colin F. "The Indian Hobbyist Movement in Europe." In *Handbook of North American Indians,* ed. William C. Sturtevant. Volume 4: *History of Indian-White Relations*, ed. Wilcomb E. Washburn, pp. 562–569. Washington, D.C.: Smithsonian Institution Press, 1988.

Vance, Lee J. "Canoeing in America." *Cosmopolitan Magazine,* October 1893, 709–716.

Wilson, G. D. "Polythene: The Early Years." In *The Development of Plastics*, ed. Susan Mossman and Peter Morris, pp. 70–86. London: Royal Society of Chemistry, 1994.

A Bulldog Travels around the World

Global Perspectives on the Tractor Production
of Heinrich Lanz and John Deere in Mannheim,
circa 1921–1965

Oliver Schmidt
Director

Westfälische Salzwelten
Bad Sassendorf, Germany

Approaches to the Globalized Agricultural Technoscape, Mannheim, Germany[1]

In 1921, the factory for agricultural machinery Heinrich Lanz, based in Mannheim, Germany, presented a new product to the market. For the first time, Lanz had developed a self-propelled tractor driven by a combustion engine and stepped away from its steam-powered traction engines of the nineteenth century, famously referred to as locomobiles.[2] Locomobiles were pulled by horses or draft cattle and used on the fields in order to support the harvesting work. Out there, they remained immobile and helped with the processes of threshing and flailing, using their steam power to run the respective agricultural engines. A belt linked them to a drive wheel that transferred the power into the threshing machine or any other helping device on the field (Figure 1). The 1921 Lanz tractor was designed to fulfill the same tasks but to be independent from any livestock. Popular narrative and memory have it that the head of the factory proclaimed "This machine looks like a bulldog" when developing engineer Fritz Huber (1881–1942) presented the

1.
Lanz locomobile. The "locomobile" was a stationary engine that could be pulled onto a field to run a variety of harvesting machines. EVZ:1991/0886 (=Bild 1997R-0157-07) TECHNO-SEUM, Photo: Klaus Luginsland.

2.
Lanz Bulldog HL 2 from 1923. The Bulldog tractor started as a self-propelling locomobile but was quickly used for all sorts of applications related to the cultivation of land. EVZ:1981/0124 (=Bild 1991R-0063-03) TECHNOSEUM, Photo: Klaus Luginsland.

world's first heavy-fuel tractor to him in the yard of the production site.[3] Ever since, the tractor was to be called, and marketed as, "Lanz Bulldog," and in several series, more than 200,000 models of it left the Mannheim works until production was terminated in 1960. The design of the tractor was obviously inspired by the bulldog image. From the front the characteristic one-cylinder, hot-bulb engine stood out, roughly resembling the head of the aforementioned dog. The front wheels added to this impression, giving the tractor the appearance of a grim and sturdy bulldog positioned in front of the viewer (Figure 2).[4] However, it was not only the appearance of this piece of machinery to which the name was supposed to refer. As a fine product of high-end agricultural technology, it was branded as "Bulldog" globally, and the attributes of the dog were to become synonymous with Lanz tractors: sturdy, indefatigable, indestructible, and full of energy. The popular and prolific design continued its mostly successful story until the 1950s, when

Lanz's decline began. In 1956 this decline was eventually halted by the takeover by John Deere, based in Moline, Illinois, currently the world's largest manufacturer of agricultural machinery. As mentioned above, Bulldog production continued until 1960, when the Bulldog product line was finally abandoned by the new masters of Mannheim's tractor works.

The TECHNOSEUM, the State Museum for Technology and Labor in Germany's southwestern state of Baden-Württemberg, conserves in its collections a series of Lanz Bulldog tractors, some of them being part of the permanent exhibition. They represent one of the three major aspects of the city's economic identity based on Mannheim's industrial past and its history of technological development and engineering. Next to Karl Friedrich Christian Ludwig von Drais (1785–1851), the inventor of the dandy horse as the forerunner of the bicycle, and Carl Benz (1844–1929), the creator of the first automobile driven by a combustion engine, stand Heinrich Lanz (1838–1905) and Fritz Huber, the corporation's arguably most significant engineer, as the names to be linked with the tractor industry of the place. Lanz became a brand that was known around the globe. The objects on display at or held in the collections of the TECHNOSEUM give some context to local industries, but inevitably, the question arises as to what degree these exhibits carry some historical meaning beyond the local, regional, and national narratives surrounding the iconic Bulldog and its manufacturer, Heinrich Lanz. As the global player John Deere joined forces with Heinrich Lanz in 1956, the step to applying a global perspective to the merging of the corporations suggests itself. Taking the 1956 watershed for Lanz as a hinge, it becomes apparent that forces of globalization worked from two directions on the fate of the plants from which the Bulldog drove into the world.

First, Lanz produced their machines for both European and global markets from the very beginning. Therefore, Lanz participated and competed in an increasingly open and globalizing arena of the agricultural market, placing their tractors in various environmental conditions and facing a range of diverse tasks with which to cope. The question I would like to pose around this complex is, How do the localized objects at the TECHNOSEUM interact with their globalized context? What narratives surround these technological artifacts in terms of their significance for the global technological development of the agricultural sector?

Second, the forces of globalization not only moved outward from Mannheim; on the contrary, with an even greater global impact, they came in. With John Deere's appearance in Mannheim, John Deere truly became a global player. The impact of the takeover allows us to put Heinrich Lanz at both the beginning and the end of a globalizing process in the agricultural machine industry. The takeover marked the end of Heinrich Lanz products but also started the wider spread of Mannheim-manufactured tractors around the world. What can the objects based at the TECHNOSEUM tell us about this significant change in the globalized industry of the manufacturing of agricultural machines? And eventually taking the journey around the globe back to Mannheim, do they tell us how these globalized artifacts become localized again, as their globality affects the circles of their production, their adaptability to their actual use at home and abroad? Finally, agriculture is probably as close to the core of globalization as possible.

Worldwide trade of agricultural products is both effect and cause of various stages of globalization.[5] To the same extent as crops have become globalized, the technology used in farming and harvesting them has become a global business. So what are the objects on agricultural technology at the TECHNOSEUM capable or incapable of telling us in respect to the globalization of their production, their application, and their users?

Heinrich Lanz's Tractor Production, 1921–1956

In the first quarter of the twentieth century, agricultural technology and machinery underwent a revolutionary change. The first petrol-driven engines appeared and were deployed in agricultural vehicles and machines alike. At the beginning of this period, motor technology consisted of locomobiles. These stationary engines and their complimentary equipment such as threshing machines were still widely used. Gradually, the predominance of steam engines in the processes of sowing, cultivation, harvesting, and the processing of the crops weakened. The first tractors and threshing machines run by internal combustion engines opened new options for greater efficiency in work procedures and in terms of employment of both human and horsepower.[6] However, skepticism toward the possibilities created by the use of combustion engines in agricultural machinery prevailed among German farmers, whereas in the United States self-propelled tractors began their triumphal march to revolutionizing agricultural productivity within the first two decades of the twentieth century.[7]

Heinrich Lanz had rather unsuccessfully started to experiment with the then called Landbaumotoren (Land Building Engines) in 1912, but German farmers criticized both the high cost of these machines in terms of purchase and maintenance and their comparatively low level of efficiency out on the field where they tended—heavy as they were—to get stuck in the soft and moist soil despite their sometimes extremely powerful engines. The Lanz Landbaumotor sported a four-cylinder, 80 hp engine weighing almost 5 tons, which made it the preferred choice for transporting heavy goods rather than cultivating land.[8] Firma Heinrich Lanz, Mannheim was only one of many enterprises that set up business in the sector of manufacturing agricultural machinery in the first half of the twentieth century. A major center of German machine works and associated industries sprang up in an area with the cornerstones of Saarbrücken to the west, Darmstadt and Würzburg to the north and the east, Augsburg in the southeast, and Freiburg in the southwest. Although far from the traditional German centers of agricultural production in the northern and eastern plains, the industrialization of the southwest combined with critical infrastructural preconditions to transform the region into a smithy of tractors, threshers, and combine harvesters. The river Rhine and the very early and comprehensive provision of railway services in the southwest provided a vast opportunity for industrial investment. By the end of the nineteenth century, the area covering today's state of Baden-Württemberg included German centers of automobile production, railway works, and a varied range of machine works and engine production.

Moreover, the southwest of Germany was by no means an agricultural wasteland. Agriculture had been a major source of income in preindustrial times. Vineyards, animal husbandry, and a limited degree of grain growing formed a stable foundation for corporations developing agricultural technology. Assembling tractors and other mobile vehicles vital to the production cycle of farms specialized in milk, cattle, crops, or wine, therefore, became an important aspect of the southwest German economy.[9] The number of long- and short-lived enterprises that produced vehicles for heavy use in agricultural production is legion in the German southwest: Agria in Möckmühl, Allgaier in Uhingen, Bautz in Saulgau, Daimler-Benz with their factory in Gaggenau, Ensinger in Michelstadt, the famous Fahr machine factory in Gottmadingen, Güldner in Aschaffenburg, Gutbrod in Saarbrücken, Heinrich Lanz in Mannheim, Hela in Aulendorf, Holder in Metzingen, Hummel in Heitersheim, Kelkel in Tamm near Ludwigsburg, Kramer in Überlingen, Krieger in Rhodt, Kühner & Berger in Sabach near Achern, Kulmus in Eisenharz, Porsche-Diesel in Friedrichshafen and Groß-Gerau, Schanzlin in Weisweil (Baden), Stihl in Waiblingen, Sulzer in Harthausen near Augsburg, Titus in Worms, Wahl in Worms, Weigold in Mannheim again, and Zanker in Tübingen—this is a list that literally spans A to Z and contains most of the corporations from southwest Germany that produced tractors and farming equipment at some point or another in the twentieth century.[10]

In this decentralized structure of agricultural industry, paradoxically, Mannheim became somewhat of a center with Heinrich Lanz—even though it was only a medium-sized enterprise— as the biggest player. Lanz's employment peaked during World War II with just over 7,000 people in their paid services, which made Lanz a large enterprise but not a global player.[11] After having started his business in 1859, importing British high technology for the German agricultural market, Heinrich Lanz began to assemble his own machines in 1867. By 1909, Heinrich Lanz had become a successful producer of steam-driven machinery. About 610,000 machines had left his factory, including 24,000 locomobiles and 15,800 complimentary threshing machines. However, the backbone of Lanz's twentieth century production became the famous Lanz Bulldog. By 1921, Lanz abandoned the steam engine business and issued the first series of Bulldogs run by a one-cylinder, two-stroke hot-bulb engine that eventually reached 55 hp in its later versions.[12]

Engineer Fritz Huber began his work on the design in 1916. It took him five years to create the first Lanz tractor equipped with a combustion engine. The 1921 HL (= Heinrich Lanz) Bulldog, however, remained fundamentally based on the idea of an engine that is driven into the field in order to provide propulsion for a range of instruments commonly used during harvest— threshers, harvesters, and similar implements. Hence, the design was, in fact, quite remarkably backward in itself, offering a solution for harvest practices of the nineteenth century. This initial lack of recognition of the potential of their new machine among the Lanz engineers could not prevent the breakthrough of the revolutionary creation they had in their hands. The customers saw this potential clearly: despite considerable difficulties they put the Bulldog to the instantaneous use of pulling machinery through their fields for plowing and for other works in the neverending process of cultivating their land.[13]

The market appearance of the Bulldog in 1921 marks the beginning of the period of motorization in the agricultural sector in Germany, which kicked off late when compared with the United States.[14] Admittedly, the phases in which the Landbaumotoren and the simultaneously backward and progressive design of tractors such as the Bulldog were constructed do much to explain the delayed impact the combustion engine would have on the agricultural sector in Germany. The first ponderous Landbaumotoren did not help to prepare the market because it needed the actual decentralized dynamic the farmers fed back to the manufacturer. This flow of information helped Lanz realize what they could do with the machines they designed. They had initially not thought of putting their tractors to the uses this term implies: to pull a range of appliances through or over a field. The first Bulldogs quickly earned themselves a reputation for their reliability, for their performance and efficiency, and, not least, for their small consumption of fuel. Additionally, Lanz soon adapted to the needs of the customers in the development of improvements and refined versions of the Bulldog.[15] The hot-bulb engine was rather difficult to ignite. It took its time to heat up, and the running of the flywheel led to quite a few injuries. The Bulldog was still a rather basic and rough model with very limited mobility—it was designed to replace the locomobiles and was, although self-propelled, very difficult to handle. Reversing the Bulldog was a near-artistic achievement that involved changing the engine's rotational direction by reducing the inflow of gas. The vibrating, loud, and mostly unprotected engine was consequently balanced in a new set of components. Over the course of the 1920s, Lanz worked on refining the suspension of its vehicles in order to achieve a more comfortable working environment for the driver.[16]

Exhibits at the TECHNOSEUM convey the technical genesis of Lanz tractors. The 1926 HR still features the drive wheel for the propulsion of threshing machines but is—as opposed to the basic Bulldog of 1921–1923—already designed to act as a proper tractor and therefore capable of pulling a variety of devices. Its 28 hp engine was quite a step forward from the mere 12 hp the original HL and HP versions provided. The developers were obviously aware of the progress they had achieved and called their new type Großbulldog (Grand Bulldog).[17] In the 1930s, Lanz introduced additional models such as the HN 2 that appealed to many customers for their simple handling. In addition, pneumatic tires were now available for the new models, which improved the range of conditions under which the machines could be used.[18] By 1939, Lanz supplied a tractor—the versatile Bauernbulldog (Farmer's Bulldog)—that incorporated such state-of-the-art features as an electric starter, an adjustable track width system, high chassis clearance, a hydraulic power lift and pneumatic-tired spoke wheels. Thus, in the interwar decades Lanz reached something commentators in Germany called "world fame,"[19] and the corporation claimed that its machines could be found all over the world and that they were suited to various needs of cultivation in different climate zones.[20] Although this claim could be substantiated by the factual presence of Lanz products on all continents with agricultural production, the extent to which they entered the market varied strongly.

Lanz had discovered foreign markets early in their history and managed to establish sales branches and consultancy offices in all the important agricultural production centers of the world such as France, the United States, India, and even Australia by 1912—long before the Bulldog appeared.[21] With the Bulldog, the corporation made some effort to adapt their tractors for competition in the global arena. For instance, Lanz's department of development closely examined to what extent materials used in a tractor faced different effects in various climatic zones. For tropical weather, Lanz used wood that was not dried to the same degree as wood that was to be used in more arid conditions; a special cooling device was mounted on HR 5 Bulldogs with the destination "tropical climate zones."[22]

In April 1939, an advertisement brochure opened with the slogan "Urwald und Steppe werden Ackerland" (Jungle and steppe become arable farmland). Below, a Lanz Bulldog tractor plowed its way through newly turned clods of earth, making the land fertile in a hostile environment. Reports from Brazil documented the indestructibility of a Bulldog tractor. It had sunk in a swamp and was eventually heaved out of it, and when heated up again, the engine started straight away, continuing its work as reliably as before the accident.[23] For Asia, Lanz reiterated the Bulldog's qualities in working under very harsh circumstances in a range of environmental conditions from forest to wasteland. Export statistics, according to Lanz, showed the Bulldog to be the most popular German tractor in the world.[24] The photographs in advertising leaflets supported this argument, depicting the Bulldog equipped with various tires and tracks used in soft, hard, swampy, or uneven ground or pulling a range of utilities and carriages. One further advantage Lanz claimed was the low cost of maintenance and operation due in part to the cheap heavy oil the hot-bulb engine burnt.[25]

The international positioning of the Bulldog was enhanced by the long-established system of field advisors Lanz maintained in Germany, in Europe, and on other continents. These traveling agents were salesmen who had been thoroughly trained to be well informed about the products' technical and commercial characteristics, the local conditions of farming, and the actual engineering of a Bulldog and its equipment.[26] They represented the axis along which customer feedback reached the developing engineers back in Mannheim.[27] Such channels of communication and direct interaction with farmers allowed the engineers to closely research the tractors' performance in action. Additionally, Lanz maintained a department that tested new vehicles and equipment out in the open field in order to react to immediate and obvious necessities of refinement. Consequently, Lanz maintained a high rank among the top innovative companies of the sector in Europe. As an example of how quickly Lanz tried to react to new opportunities on the market, just before the beginning of World War II, the corporation issued a new combine harvester that suited the harvesting requirements the recently introduced and strongly modified seed material demanded.[28]

By 1960, when John Deere stopped the product line, 219,253 Bulldog models had left the Mannheim assembly plants.[29] However, the "world fame" Lanz kept claiming for the Bulldog and complementary farming equipment appears to be a narrative based on very thin material. When Lanz celebrated the completion of the 150,000th Bulldog in 1953, only 30,000 of them had been exported—8,000 of these, almost a third of the entire export figures, went to France as the largest

European agricultural market.[30] In advertisements, Lanz reiterated the importance of exports for their business as well as the world-spanning web of sales and distribution strongpoints. Despite these announcements, Lanz remained quite silent when it came to either concrete quantities of exported products or to their destinations abroad. Quite contrary to the public branding of the Bulldog, it was not the export of this renowned machine that was to become the backbone of Lanz's economic success, but the market at home—the major part of the tractor production, 120,000 machines in total, had found its way to farms in Germany.[31]

Nevertheless, Lanz targeted the agricultural markets of Europe. In addition to France, the company sold to Spain and Ireland, seemingly with some success. Certainly, these countries did not rival France as the main export markets for Lanz, which can be explained by Mannheim's proximity to the French border. Still, Lanz's operational activity in the other European countries never sold more than several hundred or a couple of thousand machines per market. In the 1950s, Lanz claimed to export tractors and agricultural machinery to 74 countries. Leaving France's 8,000 Bulldogs out of the calculation, Lanz sold 22,000 tractors to 73 countries. Accordingly, on average, just over 300 Bulldogs reached each of them, but it must be considered that the major European markets of Spain, Ireland, and perhaps Poland until 1939 would have taken a larger share of the export. Globally, Australia, India, and probably Brazil rank among the largest buyers of Bulldogs outside Europe. A substantial or even dominant significance of the Bulldog on the global market cannot be deduced from this.[32]

Despite all rhetoric about the importance of exporting for business operation, it is a lot more likely that the global narrative was created to advertise the Bulldog at home. The tractor's features, such as high endurance, cheap maintenance, simple handling (of the later models), and versatility in all types of climates, on every soil possible, using any fuel available, related essentially to the farmers of Germany.[33] They had to be told what a fine piece of machinery they were allowed to call their property—a masterpiece of German tractor engineering that carried its name around the globe, facing up to any challenge that climate and less adept technicians abroad could possibly pose. The globalizing effects of manufacturing tractors for the agricultural markets abroad succeeded in helping to relocalize the Bulldog at home.

Eventually, this handling of both the economic and the technical development contributed largely to the end of Bulldog production and Lanz as a corporation: economically, Lanz never managed to expand to foreign markets as much as they claimed or needed for stable sales balances; technically, the adaptation to the more powerful and efficient engines that the market required came too late to boost sales of the Bulldog series. Lanz's crisis was inevitably fostered when the enterprise proved incapable of adapting to the challenges of a global market after World War II. The policy of developing, as opposed to abandoning, the design of the one-cylinder hot-bulb engine as the unique characteristic of the Bulldog emerged as insufficient to play a role in the market. Only reluctantly was the decades-old doctrine of the one-cylinder hot-bulb engine design revised when it became clearer that Lanz was trailing behind the competition and the engine simply did not fulfill the needs of the customers any longer. First semidiesel and then (but not before the mid-1950s) full diesel engines were taken into consideration for the development of new competitive models—design features global competitors such as Deere, Massey-Ferguson, and Fendt had been following to greater success since the 1920s.[34]

Simultaneously, management's lamentations about the slow business in the export markets, which supposedly caused the financial downfall of Lanz, revealed the lack of understanding of the challenges of the time.[35] Export had never been the backbone of the company's business. The acclaimed reason for success and the major trademark—the one-cylinder two-stroke hot-bulb engine Bulldog—was to become the gravestone of the enterprise. Despite record sales in 1956, the rigid adherence to this tractor design, along with the challenge of overcoming processes of production at too high a cost, paved the path to economic dependency.[36]

The Road to the John Deere Takeover and Its Effects

The wave of motorization reached the German agricultural industry after World War II with the beginning of economic growth in the early 1950s. Lanz created a wide range of products for farmers, convinced of the profits motorization would bring to their agricultural operation.[37] In 1951, Lanz introduced the tool and equipment carrier Alldog, which could take on a load of up to 750 kg.[38] The latest Lanz tractor that the TECHNOSEUM possesses is a 1954 field tractor, type D 1706. It comes with a mower and is one of the last models introduced before the John Deere takeover. Equipped with a 22 hp diesel engine, it already shows the change of development policies Lanz applied from 1952 when the so-called semidiesel engines became the new strategic component in Lanz's product line.[39] The use of this new technology helped to considerably reduce the unhealthy vibration the hot-bulb engine inevitably caused the driver. Moreover, the use of fuel was reduced by a third compared with the hot-bulb engine.[40]

However, during the years of the German "economic miracle" the competition on the market for agricultural machinery had become tighter. Many corporations entered the race to provide for the farmer's thirst for motorized vehicles and more powerful engines capable of carrying more versatile equipment. Companies such as Lanz that were so dependent on the German market were overwhelmed by both the relentless competition and the saturation of the market that was reached in the mid-1950s.[41] When John Deere took the majority of stock in Heinrich Lanz in 1956, Lanz was heavily in debt but producing better products at a greater volume than ever before. In order to adapt to new market needs, Lanz first fitted new engines in their modernized tractor series; second, the health, safety, and comfort of the customer had become ever more important in the design of new machinery. The last product line Lanz issued already reflected the driver's needs more than the more robust versions of the Bulldog in decades past. Then, the technical challenges in the field had been the greater ones. The actual submission and control of nature was the focus of the interest that customers and producers shared. In the 1950s, the European farmer not only wanted to work his land but also to feel warm, safe, and comfortable when doing so.[42]

Although these changes came too late to save Lanz as an independent company, John Deere happily continued this innovative, customer-oriented policy of research and development. By the mid-1960s, John Deere cared more about the improvement of the seat and the driver's cabin than about further increases of power or reduction of fuel usage. Moreover, the farmers were offered options to equip the basic models John Deere manufactured with a line of utilities suited to their needs. Mannheim played a major role in supplying a globalized agricultural market with these neatly adapted and innovative agricultural machines.[43]

Producing in line with the customer's requirements was one example for continuation after the takeover. Employment, however, was one aspect that changed greatly. In 1955 Lanz retained almost as many workers in their services as at the all-time high in 1940: 7,216 people worked at the Lanz works in Mannheim-Lindenhof in 1955, compared with 7,500 in 1940. In 1959, just four years later, the new leadership reduced the workforce to a mere 4,100.[44] The first four American directors of John Deere/Lanz Mannheim did not stay in service for longer than two years each. It was not until 1964 that a more stable period of consolidation would begin. The reasoning of the popular narrative read that modern American management methods and German union traditions clashed and stalled the recovery of Lanz's enterprise for years.[45] In fact, the contrary happened. The new management aimed at modernizing both the manufacturing process and the deployment of the workforce. This meant specifically rationalizing the production line and reassigning the workers, specialist craftsmen, and engineers in a more efficient fashion.[46] John Deere tried to adjust the Mannheim works to the challenges of the global market as opposed to the regional direction the Lanz works had followed until 1956. The Lanz product lines were allowed to run out. Simultaneously, management prepared the factories to take on the assembly of the new John Deere series of tractors and agricultural equipment. When the old Lanz production was terminated in the early 1960s and the John Deere machines started to enter the markets, the recovery of the Mannheim complex came within a few years. By the end of the 1960s, Lanz-Deere works in Mannheim were as healthy as Lanz had probably been only before the financial crisis of 1929.[47]

Although John Deere initially maintained the production of the Bulldog only to terminate it in 1960, John Deere had managed to buy their way into the European market. The technically old-fashioned Bulldog was still an established brand, mostly in Germany but in Europe, too. The continuation of sales helped raise expectations for new models and prepared their introduction.[48] With the new stronghold in central Europe, the American corporation divided production and development according to respective market needs. Deere kept the assembly of its more powerful machines at the parent plant in Moline. Expressing some appreciation for German engineering in general and Lanz's development history in particular, the model lines 300 and 500, less powerful but better suited to the major European agricultural markets in France and Germany, were given to Mannheim, where they were independently designed according to the wishes expressed by the headquarters.[49] Launched with a massive campaign in 1960, the former Lanz works introduced once more a completely innovative line of products into the agricultural technoscape. Technical devices and additions, apart from the four-cylinder and four-stroke Diesel engine, such as a ten–gear range gearbox, revolutionary hydraulic systems, and disc brakes along with improved comfort for the driver, set completely new standards for agricultural machinery in Europe and introduced a new range of expectations of what a tractor had to be capable of.[50] In effect, the John Deere takeover changed more than the color of the machines, which were altered from Lanz blue to Deere green. By the mid-1960s the change of production, development, and redeployment of the workforce was complete, and the business figures started to show improvement again. Formerly the largest producer of tractors in Germany, Lanz was now part of the largest producer of agricultural utility vehicles in the world.[51] John Deere managed to combine their own global infrastructure with the finely spun web of Lanz consultancy offices on all continents with agrarian production.[52]

To a certain extent, John Deere's engagement in Europe after World War II therefore follows the model of a typical globalizing American multinational. The market remained difficult to access even after American economic power in Europe had continued to grow after the end of the war. The takeover of a well-established and influential company granted the opportunity to participate in European agricultural development—a large business as the wave of motorization still continued throughout the 1950s, roughly 30 years after the motorization of the American agricultural industry.[53]

As a consequence, the TECHNOSEUM's MD 18 S combine harvester, built from 1956 onward, represents the end and the beginning of an era.[54] Although it made steam power as well as a range of related machinery redundant, it embodied the arrival of the era of motorization in German agriculture in the 1950s. It also represents the end of the history of Heinrich Lanz as an independent enterprise, as the new combine harvester was already designed in the colors of John Deere products—abandoning the traditional design of Lanz models. It also represents yet another aspect of globalization: the restructuring of enterprise complexes due to the global orientation of internationally operating groups. The impact of John Deere on Lanz works in Mannheim extended beyond the mere replacement of production lines and rationalizing of both work processes and employment. The whole production of combine harvesters, for instance, was ultimately moved to Zweibrücken, roughly 100 kilometers west of Mannheim.[55] The globalizing effect the takeover showed here was that John Deere gave a new, clearly defined structure to the Lanz complex of machine works and simply assigned precisely formulated tasks to the respective factories. Mannheim had to produce tractors, Zweibrücken had to produce combine harvesters, and the headquarters in Mannheim had to organize sales and distribution for both works and for the areas the Moline-based group headquarters designated for Lanz business.[56] By introducing and establishing their presence in the European market, John Deere had truly become a global player, now also dividing not only production lines but also the world according to what works distributed which products to the respective markets. The U.S. headquarters initially reserved the Americas, Australia, and Asia for themselves, assigning Mannheim the task of producing for Europe, Africa, and the Middle East—successfully. What Lanz had claimed for decades, Deere realized: by 1970 of 15,631 sales, 78% were exported, and only 22% stayed in Germany. The import-export ratio had been reversed. Although Germany's agricultural market was targeted by Deere, the new manufacturer explicitly modeled their business in Mannheim for specific export markets. The overall numbers of sales increased, and the proportion of product that stayed in Germany declined in a market that was highly contested.[57]

Conclusion

Having traveled around the world with Heinrich Lanz and their famous Lanz Bulldog and returned to Mannheim, Germany, via Moline, Illinois, and having settled here again with John Deere, what have the objects surrounding these enterprises and, most of all, the Lanz Bulldog told us in respect to the globalization of agriculture, its machinery, and its production? Examining the actual artifacts and their individual history, it has to be stated that the various tractors in question were built in Mannheim and used in regional or, at most, national circumstances—none of them has ever left the area. Even though it would have been interesting to tell the story of how an individual object traveled to India or Australia, where local farmers adapted the machine to their respective needs, and then, via whatever obscure routes, found its way back to some relative in Germany or Europe, where further changes were applied or removed, and how the tractor was then eventually incorporated in the TECHNOSEUM's collection, hopes like this are seldom fulfilled. To the contrary, historians of technology will have to accept that ideal research situations related to object inventories are very unlikely. Therefore, it cannot be the objects' individual histories whose analyses bear the fruit of knowledge, only the contextualization of the objects with respect to the questions asked about them.[58]

When looked at from a global viewpoint, the Bulldog, for instance, reveals its completely national narrative. Despite being marketed on all continents with considerable agricultural activity, it remained a localized product that was endowed with a global narrative to make it more attractive to German farmers who were convinced of the brilliance of the engineering that their vehicle embodied and that spanned the globe. Creating a global web of an agricultural technoscape, the farmers were capable of relating their work to that of farmers in India, France, and South Africa, although shared experience leads to shared narrative only to a certain extent. The global narrative also reimported the tale of German technical prowess and engineering genius. In many parts of the world, farmers needed close technical supervision and training when dealing with the Bulldog, which sometimes caused even German farmers quite a few difficulties when trying to remain uninjured while handling it.[59]

However, the principal backwardness of the Bulldog's design also tells the story of the global failure experienced by Germany's then largest manufacturer of tractors. The Bulldog's story is therefore one of ambiguous hybridity and imagination. The seemingly great success story of a revolutionary design is rather a story against all odds and not nearly as revolutionary as related by its engineers. Difficult to handle, not initially designed for purposes it was used for by the consumers, and equipped from the beginning with technology that was not entirely competitive, it turns out to be rather unlikely that Bulldog became known globally as a name for a product from Mannheim. However, on the German side and with regard to the actual success in its home country, the imagination of the global narrative contributed enormously to the establishment of the tale of the Bulldog.

Notes

1. On the concept of the "technoscape," see Arjun Appadurai, "Disjuncture and Difference in the Global Cultural Economy," *Theory, Culture and Society* 7 (1990): 297–298.

2. Actually, from 1912 Lanz built the Lanz Landbaumotor (Land Building Engine), a tractor of massive measurements and 80 hp weighing roughly 4,800 kg. However, Lanz built the machine under license, and it had been developed from 1908 by a Hungarian engineer. The 1921 product was Lanz's own design. Moreover, the Landbaumotor's production was terminated as early as 1926; the Landbaumotor was used only in large agricultural operations and, in fact, became more famous for having been used by the Imperial German Army during World War I as an artillery transporter. Transport was to become the major use of the machine—less useful for cultivating land, it was capable of transporting great loads. See Udo Paulitz, *Deutsche Traktoren, 1920–1970: Von Allgaier bis Zettelmeyer* (Königswinter, Germany: Heel Verlag, 2002), 118–119. As a general guideline for the purpose of easier readability, I will refer to the mentioned steam-powered traction engines as locomobiles henceforth.

3. Other stories read that it was the workers who assigned the name "Bulldog" to the tractor: Wolfgang Wagner, *Als der Bulldog grün wurde: Traktorengeschichte 1945–1967* (Frankfurt am Main: DLG-Verlags, 2005), 22. Certainly, the merit of inventing the tractor's popular name cannot easily be ascribed to a single person.

4. See Paulitz, *Deutsche Traktoren*, 119; Gerhard Zweckbronner, "Mechanisierung der Landarbeit: Der Lanz-Bulldog im landwirtschaftlichen Technisierungsprozeß," in *Räder, Autos und Traktoren: Erfindungen aus Mannheim; Wegbereiter der mobilen Gesellschaft* (Mannheim, Germany: Landesmuseum für Technik und Arbeit, 1986), 103, 110. For the visualization of the tractor in its resemblance to a bulldog, see also Wagner, *Traktorengeschichte 1945–1967*, 16. One of Lanz's greatest rivals, Klöckner-Humboldt-Deutz (KHD), issued its first creation of a self-propelled tractor five whole years later in 1926. However, KHD fitted its machines with diesel engines from the very beginning; see Gustav Goldbeck, *Kraft für die Welt: 1864–1964 Klöckner-Humboldt-Deutz AG* (Düsseldorf: Econ-Verlag, 1964), 172–173.

5. On globalization, see, generally, Martin Albrow, "Globalization," in *The Blackwell Dictionary of Twentieth Century Social Thought*, ed. William Outhwaite and Tom Bottomore (Cambridge, MA: Blackwell, 1993), 248–249; and specifically on the economic and cultural impact of globalization processes, Bruce Mazlish, "Global History and World History," in *The Global History Reader*, ed. Bruce Mazlish and Akira Iriye (New York: Routledge, 2005), 16–20; Michael Geyer and Charles Bright, "World History in a Global Age," in Mazlish and Iriye, *The Global History Reader*, 21–29; Arjun Appadurai, "Cultural Dimensions of Globalization," in Mazlish and Iriye, *The Global History Reader*, 276–284; Anthony Giddens, "The Globalizing of Modernity," in Mazlish and Iriye, *The Global History Reader*, 285–291 (New York: Routledge, 2005).

6. For threshing machines in Germany, see Willi Plöchl, *Dreschmaschinen: Geschichte, Entwicklung, Technik, Prospekte* (Hahnstätten-Zollhaus: Bulldog Press, 1996).

7. Hans Jürgen Mathies, ed., *Die Entwicklung des landwirtschaftlichen Maschinenwesens in Deutschland* (1910; repr., Düsseldorf: VDI-Verlag, 1987), 51–52. For the development of tractors in the United States, see Sigfried Giedion, *Die Herrschaft der Mechanisierung: Ein Beitrag zur anonymen Geschichte* (1948; repr., Frankfurt am Main: Europäische Verlagsanstalt, 1982), 189–190, 279.

8. For the appearance of Landbaumotoren as the first tractors in Germany in the first decade of the twentieth century, see Mathies, *Landwirtschaftlichen Maschinenwesens*, 51–52. Almost simultaneously, some of Lanz's German competitors designed similar models that remained similarly unsuccessful, such as the World War I era Deutzer Trecker by KHD; Goldbeck, *Kraft für die Welt*, 172. For Lanz's Landbaumotor, see also Wagner, *Traktorengeschichte 1945–1967*, 103.

9. On the industrial development of the southwest of Germany, see Thomas Vogel, *Wirtschaftswundermusterländle: Baden und Württemberg in den 50er, 60er und 70er Jahren* (Stuttgart: Theiss, 2006); Alfred E. Ott, ed., *Die Wirtschaft des Landes Baden-Württemberg* (Stuttgart: Kohlhammer, 1983).

10. For more detailed articles on the individual companies, see Klaus Herrmann, *Traktoren in Deutschland 1907 bis heute: Firmen und Fabrikate* (Munich: BLV Verlagsgesellschaft, 2000); for another and more current overview, see Georg Bauer, *Faszination Landtechnik: 100 Jahre Landtechnik; Firmen und Fabrikate im Wandel* (Frankfurt am Main: DLG-Verlag, 2003). For Allgaier, see also the short film "Allgaier-Porsche A 111," Original-Allgaier-Film, Stuttgart, 1954. The TECHNOSEUM also holds an Allgaier tractor: TECHNOSEUM, EVZ:1988/1290: Schlepper Allgaier, 1950. This list of corporations also reflects the economic structure of the southwest, featuring many small- and medium-sized production facilities spread over the country in a decentralized fashion. Although major industrial centers such as Munich, Nuremberg, Berlin, Hanover, and Cologne also housed some of the largest and most successful players in the development of agricultural technology, in the southwest especially, many corporations in remote areas found their niches in the market. Eventually, global players such as Fiat and Lamborghini entered the German market after World War II with production sites in Heilbronn and Groß-Gerau, respectively. For a history of the major player and Lanz competitor MAN, see Johannes Bähr, Ralf Banken, and Thomas Flemming, *Die MAN: Eine deutsche Industriegeschichte* (Munich: C. H. Beck, 2008); and for KHD/Deutz in Cologne, see Goldbeck, *Kraft für die Welt*.

11. As a comparison between Lanz as machine works focusing on agricultural technology and the more diversified global player MAN operating in agriculture, mobilization, and machine works alike, employment at MAN grew from just below 10,000 in 1908 to over 18,000 in 1950 and 35,000 in 1967; Bähr et al., *MAN*, 359, 380, 561. For Lanz, see Udo Paulitz, *Deutsche Traktoren 2,*

1920–1970: Von Aktivist bis Zanker (Königswinter, Germany: Heel Verlag, 2003), 134–135. The central position Lanz took in the southwest of Germany was favored by the coincidental development of industries complimentary to Lanz's tractor production in Mannheim and its surroundings. One of these companies was Motorenwerke Mannheim (Engine works Mannheim; MWM), which eventually became one of the two most important providers of engines in tractors and similar vehicles in Germany. Particularly in the early 1950s, when the German motorization of the agricultural industry boomed, MWM provided the engines for a variety of short-lived production series as well as successful and enduring production lines, such as those of Kramer (Überlingen) and Norddeutsche Traktorenfabrik (Nordtrak) based in Hamburg. Another example of a corporation that developed components for the production of agricultural utility vehicles is Zahnradpumpenfabrik Mannheim (Gear pump factory Mannheim; ZPM), which began production in 1904. In these machine works pumps for oil, diesel engines, and lubricants were developed; Friedhelm Meier, *Einhundert Jahre für die Landtechnik-Industrie: Die Geschichte des Verbandes; Vom Verein der Fabrikanten zur LAV* (Frankfurt am Main: MaschinenbauVerlag, 1997), 142–146; Herrmann, *Traktoren*, 27, 118, 147.

12. For a good overview of the whole selection of Lanz agricultural machinery before World War II, see *Ein Gang durch die Lanz-Werke* (1937; repr., Obershagen, Germany: Schwungrad-Verlag, 1993), as well as Kurt Häfner and Michael Karle, *Lanz: Landmaschinen-Prospekte von 1935 bis 1945* (Stuttgart: Kosmos, 1998).

13. Paulitz, *Deutsche Traktoren*, 120–122.

14. For the impact of the tractor in the United States and in Germany, see Frank Uekötter, *Die Wahrheit ist auf dem Feld: Eine Wissensgeschichte der deutschen Landwirtschaft* (Göttingen: Vandenhoek & Ruprecht, 2010), 277–279.

15. *Ein Gang durch die Lanz-Werke*, no page. In fact, Lanz was one of the first companies to establish a comprehensive training service for its customers; Uekötter, *Wissensgeschichte*, 292.

16. For difficulties in the handling of the Bulldog and the solutions Lanz offered, see Herrmann, *Traktoren*, 126–130; Paulitz, *Deutsche Traktoren*, 119–126.

17. Available at the TECHNOSEUM as an ensemble: EVZ:1986/0193; 1986/0193-001; 1986/0193-002, consisting of the Moorbulldog-Pfluggespann, 1930–1939, and Schlepper: Lanz Bulldog, 1937.

18. Herrmann, *Traktoren*, 126–127. A 1937 Moorbulldog equipped with a plow and a very impressive set of moor tires shows what answers Lanz found for a variety of challenging tasks in everyday agricultural work. TECHNOSEUM, EVZ:1981/0124, Bulldog: HR 2, 1926; Herrmann, *Traktoren*, 128–130; John Deere Werke Mannheim, *Geschichte der John Deere Werke Mannheim* (Mannheim: John Deere, 1979), 23.

19. Max Hofer, *Heinrich Lanz A. G. Mannheim* (Berlin: Organisation Verlagsgesellschaft, 1929), 83; *Ein Gang durch die Lanz-Werke*, no page.

20. For the significance that these challenges arising from the need to adapt the tractors to various climate zones had for the development at Heinrich Lanz and for the requirements made of the material used in production, see Hofer, *Heinrich Lanz*, 26, 52–53.

21. *Ein Gang durch die Lanz-Werke*, no page.

22. Hofer, *Heinrich Lanz*, 67; John Deere Werke Mannheim, *John Deere Werke*, 24.

23. For another example of stories like this, also from Brazil and about yet another overturned Bulldog, see *Ein Gang durch die Lanz-Werke*, no page. For the war era models and technical developments, see Kurt Häfner, *Lanz: Holzgas-, Raupen-, Nachkriegs-Bulldogs von 1942 bis 1955*. Stuttgart: Franckh-Kosmos, 1990.

24. For the global trajectory of Lanz long before World War II, see *Ein Gang durch die Lanz-Werke*, no page.

25. For the leaflet, see John Deere Archiv, Mannheim, cupboard 3, drawer 1, file A.

26. For the traveling agents, see John Deere Archiv, cupboard 1, drawer 2, CI/II 20. For the written version of advice and the instructions for the salesmen, see cupboard 7, drawer 3.

27. Hofer, *Heinrich Lanz*, 79–81, 83.

28. *Ein Gang durch die Lanz-Werke*, no page.

29. Bauer, *Landtechnik*, 227; Kurt Häfner, *Der Lanz Bulldog von 1952 bis 1962* (Schwieberdingen: Verlag Klaus Rabe, 1986), 138–139.

30. John Deere Werke Mannheim, *John Deere Werke*, 31; Wagner, *Traktorengeschichte 1945–1967*, 62–63.

31. Wagner, *Traktorengeschichte 1945–1967*, 47, 72. In 1952–1953 Lanz claimed the top position among German tractor producers.

32. On Lanz's export see Wagner, *Traktorengeschichte 1945–1967*, 46–47, 54–55, 124–125. The Netherlands and Greece may also be mentioned as larger markets for Lanz's sales. Because of the dawn of the Cold War, eastern European countries were not open to Lanz's products. In 1951, the internal works' newspaper *Der Lanz Turm* claimed that exports were going to almost 40 countries again and somewhat blurrily that Lanz was performing well to help German exports increase to prewar levels. Simultaneously, the newspaper blamed foreign import restrictions for stalling trade and lamented the stiff competition the enterprise was facing abroad. Two years later, Lanz celebrated the 150,000th Bulldog. Lanz tractors were then exported to 74 countries. Wagner, *Traktorengeschichte 1945–1967*, 104–105.

33. John Deere Werke Mannheim, *John Deere Werke*, 33.

34. For reactions to international competitors, see Wagner, *Traktorengeschichte 1945–1967*, 46–47; for the development at Lanz works, see John Deere Werke Mannheim, *John Deere Werke*, 30.

35. Wagner, *Traktorengeschichte 1945–1967*, 46–47.

36. For a more detailed description of the technical performance of a wider range of Lanz models, see Paulitz, *Deutsche Traktoren*, 118–149. For the technical change in the course of the 1950s and, consequently, after the John Deere takeover, see Kurt Häfner, *Lanz: Halbdiesel-, Volldiesel-Bulldogs, John-Deere-LANZ-Schlepper von 1952–1967* (Stuttgart: Franckh-Kosmos, 1991).

37. Herrmann, *Traktoren*, 128–129.

38. The Alldog resembled the American military truck design of the M274 Mechanical Mule, which was deployed especially during the Vietnam War with a similar purpose: the transport of tools or, more likely in the M274's case, weapons. For details about the Alldog, see Kurt Häfner, *Lanz: Alldog-Geräteträger von 1951 bis 1960* (Stuttgart: Franckh-Kosmos, 1993). The TECHNO-SEUM is in possession of an Alldog model, too: TECHNOSEUM, EVZ:1989/0409, Traktor: Alldog Heinrich Lanz, 1954. Additionally, the archives of the TECHNOSEUM contain a variety of commercial material surrounding the market implementation of this new Lanz model: TECHNOSEUM, PVZ:1992/R-0160/5-12, Werbeprospekt: Lanz-Alldog, n.a.; PVZ:1992/R-0160/5-11; PVZ:1992/R-0160/5-07; PVZ:1992/R-0160/5-09; PVZ:1992/R-0160/5-07; PVZ:1992/R-0160/5-05; PVZ:1992/R-0160/5-03, and PVZ:1992/R-0160/5-01, all: Prospekt: Lanz-Alldog, n.a. The Alldog could also be utilized in pest control; Martin Zimmermann, *Schlipf: Praktisches Handbuch der Landwirtschaft*, 32nd ed. (Hamburg: Verlag Paul Parey, 1958), 133.

39. For this new development at Lanz and the D 1706, especially designed for the export market, see Häfner, *Lanz Bulldog*, 19–23, 28–29. The TECHNOSEUM's model is available as EVZ:1989/0462, Ackerschlepper: D 1706, 1954, and, in fact, is on display on level F of the State Museum.

40. Herrmann, *Traktoren*, 130.

41. For the decrease in exports and market saturation in the 1950s, see Wagner, *Traktorengeschichte 1945–1967*, 166–167.

42. Bauer, *Landtechnik*, 226–227.

43. Herrmann, *Traktoren*, 42–43. For a more detailed narrative of the merging of the two companies, see Don Macmillan, *John Deere: Die Geschichte des größten Traktorenherstellers der Welt* (Königswinter, Germany: Heel Verlag, 2010), 258–261. For attempts to stay on top of innovation, see Uekötter, *Wissensgeschichte*, 311.

44. John Deere Werke Mannheim, *John Deere Werke*, 32–35, 52. For a comparison to national rivals, KHD employed more than 9,000 workers in 1949. With a similar reduction of staff when the market was saturated between 1956 and 1958, in the early 1960s, KHD grew to 23,490 workers plus 9,089 employees; Goldbeck, *Kraft für die Welt*, 275.

45. John Deere Werke Mannheim, *John Deere Werke*, appendix "Executives."

46. On the comprehensive endeavor of modernizing the plant lasting until 1967, see John Deere Werke Mannheim, *John Deere Werke*, 39.

47. On the modernization of the Mannheim works, see Wagner, *Traktorengeschichte 1945–1967*, 248–249.

48. For the parallel manufacturing of the product generations, see Wagner, *Traktorengeschichte 1945–1967*, 176–177.

49. Wagner, *Traktorengeschichte 1945–1967*, 191–195.

50. For the introduction of the new Deere tractor series, see Wagner, *Traktorengeschichte 1945–1967*, 255. The new Deere series was referred to as the "10er line." See also Macmillan, *John Deere*, 260–261.

51. John Deere Werke Mannheim, *John Deere Werke*, 36–39, 51–52; Bauer, *Landtechnik*, 80–81; Herrmann, *Traktoren*, 40–42.

52. On Lanz's well-organized network of customer service facilities, see Wagner, *Traktorengeschichte 1945–1967*, 117.

53. For a model of American multinationals' development, see Mira Wilkins, "Mapping Multinationals," in Mazlish and Iriye, *The Global History Reader*, 80–81.

54. For the technical development of combine harvesters in Germany after World War II, see Alfons Eggert, *Von der Mähmaschine zum Mähdrescher: Die Technik in der Getreideernte* (Münster: Aschendorff Verlag, 1991), 94–99; for the MD 18 S specifically, see Wagner, *Traktorengeschichte 1945–1967*, 228–232. The first fully automatic and self-propelled combine harvester was developed in the United States by Massey-Ferguson as early as 1936.

55. For the restructuring process, see Wagner, *Traktorengeschichte 1945–1967*, 232–233.

56. John Deere Werke Mannheim, *John Deere Werke*, 38–39.

57. John Deere Werke Mannheim, *John Deere Werke*, 41.

58. On the concept of "biographies of things," see Igor Kopytoff, "The Cultural Biography of Things: Commoditization as Process," in *The Social Life of Things: Commoditization in Cultural Perspective*, ed. Arjun Appadurai (Cambridge, MA: Cambridge University Press, 1986), 66–68. More suitable with respect to the objects under consideration is the concept of "thickening of significance" by adding both context and the cultural imagination and discourse surrounding the artifacts to the examination in order to make them talk. For this concept, see Lorraine Daston, "Speechless," in *Things That Talk: Object Lessons from Art and Science*, ed. Lorraine Daston (New York: Zone Books, 2004), 18–20, 24. Complementary to this complex, see also Arjun Appadurai, "Globale ethnische Räume: Bemerkungen und Fragen zur Entwicklung einer transnationalen Anthropologie," in *Perspektiven der Weltgesellschaft*, ed. Ulrich Beck (Frankfurt am Main: Suhrkamp, 1998), 17–18.

59. For technical training abroad, especially in India and South Africa, as well as for the self-perception of German technological superiority, see Wagner, *Traktorengeschichte 1945–1967*, 54–55, 72–73, 106–107, 220–221.

Bibliography

Albrow, Martin. "Globalization." In *The Blackwell Dictionary of Twentieth Century Social Thought*, ed. William Outhwaite and Thomas Bottomore, pp. 248–249. Cambridge, MA: Blackwell, 1993.

"Allgaier-Porsche A 111." Original-Allgaier-Film. Stuttgart, 1954.

Appadurai, Arjun. "Cultural Dimensions of Globalization." In *The Global History Reader*, ed. Bruce Mazlish and Akira Iriye, pp. 276–284. New York: Routledge, 2005.

———. "Disjuncture and Difference in the Global Cultural Economy." *Theory, Culture and Society* 7 (1990): 295–310.

———. "Globale ethnische Räume: Bemerkungen und Fragen zur Entwicklung einer transnationalen Anthropologie." In *Perspektiven der Weltgesellschaft*, ed. Ulrich Beck, pp. 11–40. Frankfurt am Main: Suhrkamp, 1998.

Bähr, Johannes, Ralf Banken, and Thomas Flemming. *Die MAN: Eine deutsche Industriegeschichte*. Munich: C. H. Beck, 2008.

Bauer, Georg. *Faszination Landtechnik: 100 Jahre Landtechnik; Firmen und Fabrikate im Wandel*. Frankfurt am Main: DLG-Verlag, 2003.

Daston, Lorraine. "Speechless." In *Things That Talk: Object Lessons from Art and Science*, ed. Lorraine Daston, pp. 9–24. New York: Zone Books, 2004.

Eggert, Alfons. *Von der Mähmaschine zum Mähdrescher: Die Technik in der Getreideernte*. Münster: Aschendorff Verlag, 1991.

Ein Gang durch die Lanz-Werke. 1937. Reprint, Obershagen, Germany: Schwungrad-Verlag, 1993.

Geyer, Michael, and Charles Bright. "World History in a Global Age." In *The Global History Reader*, ed. Bruce Mazlish and Akira Iriye, pp. 21–29. New York: Routledge, 2005.

Giddens, Anthony. "The Globalizing of Modernity." In *The Global History Reader*, ed. Bruce Mazlish and Akira Iriye, pp. 285–291. New York: Routledge, 2005.

Giedion, Sigfried. *Die Herrschaft der Mechanisierung: Ein Beitrag zur anonymen Geschichte*. 1948. Reprint, Frankfurt am Main: Europäische Verlagsanstalt, 1982.

Goldbeck, Gustav. *Kraft für die Welt: 1864–1964 Klöckner-Humboldt-Deutz AG*. Düsseldorf: Econ-Verlag, 1964.

Häfner, Kurt. *Der Lanz Bulldog von 1952 bis 1962*. Schwieberdingen: Verlag Klaus Rabe, 1986.

———. *Lanz: Alldog-Geräteträger von 1951 bis 1960*. Stuttgart: Franckh-Kosmos, 1993.

———. *Lanz: Halbdiesel-, Volldiesel-Bulldogs, John-Deere-LANZ-Schlepper von 1952–1967*. Stuttgart: Franckh-Kosmos, 1991.

———. *Lanz: Holzgas-, Raupen-, Nachkriegs-Bulldogs von 1942 bis 1955*. Stuttgart: Franckh-Kosmos, 1990.

Häfner, Kurt, and Michael Karle. *Lanz: Landmaschinen-Prospekte von 1935 bis 1945*. Stuttgart: Kosmos, 1998.

Herrmann, Klaus. *Traktoren in Deutschland 1907 bis heute: Firmen und Fabrikate*. Munich: BLV Verlagsgesellschaft, 2000.

Hofer, Max. *Heinrich Lanz A. G. Mannheim*. Berlin: Organisation Verlagsgesellschaft, 1929.

John Deere Archiv, Mannheim.

John Deere Werke Mannheim. *Geschichte der John Deere Werke Mannheim*. Mannheim: John Deere, 1979.

Kopytoff, Igor. "The Cultural Biography of Things: Commoditization as Process." In *The Social Life of Things: Commoditization in Cultural Perspective*, ed. Arjun Appadurai, pp. 64–91. Cambridge, MA: Cambridge University Press, 1986.

Macmillan, Don. *John Deere: Die Geschichte des größten Traktorenherstellers der Welt*. Königswinter, Germany: Heel Verlag, 2010.

Mathies, Hans Jürgen, ed. *Die Entwicklung des landwirtschaftlichen Maschinenwesens in Deutschland*. 1910. Reprint, Düsseldorf: VDI-Verlag, 1987.

Mazlish, Bruce. "Global History and World History." In *The Global History Reader*, ed. Bruce Mazlish and Akira Iriye, pp. 16–20. New York: Routledge, 2005.

Meier, Friedhelm. *Einhundert Jahre für die Landtechnik-Industrie: Die Geschichte des Verbandes; Vom Verein der Fabrikanten zur LAV*. Frankfurt am Main: MaschinenbauVerlag, 1997.

Ott, Alfred E., ed. *Die Wirtschaft des Landes Baden-Württemberg*. Stuttgart: Kohlhammer, 1983.

Paulitz, Udo. *Deutsche Traktoren, 1920–1970: Von Allgaier bis Zettelmeyer*. Königswinter, Germany: Heel Verlag, 2002.

———. *Deutsche Traktoren 2, 1920–1970: Von Aktivist bis Zanker*. Königswinter, Germany: Heel Verlag, 2003.

Plöchl, Willi. *Dreschmaschinen: Geschichte, Entwicklung, Technik, Prospekte*. Hahnstätten-Zollhaus: Bulldog Press, 1996.

Uekötter, Frank. *Die Wahrheit ist auf dem Feld: Eine Wissensgeschichte der deutschen Landwirtschaft*. Göttingen: Vandenhoek & Ruprecht, 2010.

Vogel, Thomas. *Wirtschaftswundermusterländle: Baden und Württemberg in den 50er, 60er und 70er Jahren*. Stuttgart: Theiss, 2006.

Wagner, Wolfgang. *Als der Bulldog grün wurde: Traktorengeschichte 1945–1967*. Frankfurt am Main: DLG-Verlags, 2005.

Wilkins, Mira. "Mapping Multinationals." In *The Global History Reader*, ed. Bruce Mazlish and Akira Iriye, pp. 79–89. New York: Routledge, 2005.

Zimmermann, Martin. *Schlipf: Praktisches Handbuch der Landwirtschaft*. 32nd ed. Hamburg: Verlag Paul Parey, 1958.

Zweckbronner, Gerhard. "Mechanisierung der Landarbeit: Der Lanz-Bulldog im landwirtschaftlichen Technisierungsprozeß." In *Räder, Autos und Traktoren: Erfindungen aus Mannheim; Wegbereiter der mobilen Gesellschaft*, pp. 96–115. Mannheim, Germany: Landesmuseum für Technik und Arbeit, 1986.

CHAPTER 10

Technology Heritage Online
A Review of the Digital Museum
Inventing Europe

Kimberly Coulter

*Director of the Environment &
Society Portal*

*The Rachel Carson Center for
Environment and Society*
LMU-Munich
Munich, Germany

*Digital projects—like all technological artifacts—
may shape,* and be shaped by, diverse knowledge
cultures. This chapter reviews a digital project that
is remarkable not only for its integration of objects
from diverse European collections but also for its
creators' awareness and conceptualization of the
project itself as a technological artifact. The "digital
museum" *Inventing Europe: European Digital Museum for Science and Technology* launched in April
2013 with the goal of generating "critical historical reflection on the prominent technological processes and narratives of European integration as
technological progress," according to its former editor in chief, Alexander Badenoch. "As such,"
he explains, "the collaborative online platform for circulating (digital) artefacts and knowledge
is inevitably enmeshed in the very sort of processes it seeks to explore."[1] In this review, I assess
how *Inventing Europe* works as a digital project—mobilizing structure, design, and tools for dissemination—to achieve this objective.

In addition to disciplinary expertise, creating a good digital project requires three qualities:
critical understanding of how technologies affect knowledge creation and diffusion, the ability to
work with information from different cultures and infrastructures, and effective communication
with diverse partners and audiences. These are challenges even for historians who work within
one national, institutional, or disciplinary context. But the reward is great: allowing users to

locate and access sources more readily than ever before; to view, compare, and analyze them in new ways; and to share the resulting stories with broad publics. As Todd Pressner observes in his "Digital Humanities Manifesto,"

> Universities—no longer the sole producers, stewards, and disseminators of knowledge or culture—are called upon to shape natively digital models of scholarly discourse for the newly emergent public spheres of the present era (the www, the blogosphere, digital libraries, etc.), to model excellence and innovation in these domains, and to facilitate the formation of networks of knowledge production, exchange, and dissemination that are, at once, global and local.[2]

The in situ museum offers visitors the opportunity to physically immerse themselves in its collections. Wandering through the Deutsches Museum, the visitor beholds all manner of objects from the humble canning pot to the formidable airplane—objects able to engage imaginations and memories in ways photographs cannot. Online exhibitions, however, can virtually juxtapose objects that otherwise would not come together physically. Texts, metadata, detail views, and links make it possible to research an object at any time, without traveling to the museum. Digital projects thus not only extend the traditional museum's audience; they connect these users to a potentially unlimited number of related resources. Accordingly, they hold the potential to expand—even transform—objects' narratives.

Global Narratives, Local Objects

As an online project, *Inventing Europe* emerges from a larger research agenda led by the history of technology research network Tensions of Europe. The network aims to advance alternative understandings of European integration which, in lieu of a dominant focus on state policies and institutions, take a view of technology in a global context as providing a kind of "integration from below," paying attention to ways technology influences and is influenced by local practices. The digital project is related to a slate of four projects, funded from 2007 to 2010 under the European Science Foundation's EUROCORES scheme, and the *Making Europe* book series (2013–2014). The project's early prototype, *Europe, Interrupted*, aimed to support networking within this research context and disseminate the results to the broader public. Under the direction of project leader Johan Schot, now director of the Science Policy Research Unit of the University of Sussex, the redesigned *Inventing Europe* site employs a thematic structure corresponding to the *Making Europe* book series, a user-friendly interface, and the new Europeana API (application programming interface) 2.0, which feeds related examples from the Europeana Foundation's massive collection of 26 million items.[3] The site builds (virtual) material, institutional, and discursive connections—both a presentation of the network's results as well as an example of European integration in practice.

Three levels of technology narratives drive the project's structure. "Exhibitions" define the project's thematic scope, "stories" provide the building blocks, and the story-assembling "tours"

1.
Homepage of *Inventing Europe: European Digital Museum for Science and Technology* (http://www.inventingeurope.eu).
Photo: Foundation for the History of Technology, Eindhoven (the Netherlands). Accessed 22 August 2014.

serve as the user's primary mode of engagement with—and perhaps one day content generation for—the site. Upon entering the site *Inventing Europe*, visitors choose from six exhibitions
(Figure 1).

The *Daily Lives* exhibition contextualizes objects that have shaped home economics and
lifestyles over the last 150 years. Users can peruse European objects related to transport, communication, and energy in the *Infrastructures* exhibition and learn about the political issues of
technology coordination and standards development in *Governance*. Global effects—such as political revolutions—of communication technologies from the telegraph to satellites are discussed
in the *Communications* exhibition. *Globalisation* focuses on the presentation of Europe at world's

fairs, women adventurers exploring the world, and the infrastructures connecting Europe with colonies and migrants. Finally, *Knowledge Societies* presents objects related to scientific training and the often politicized use of scientific knowledge.

The site's logic is driven by individual cultural heritage objects themselves; it does not aim to offer a comprehensive overview of European technology or technology's significance for the European project. The heart of the site's 200-some microstories are cultural heritage objects: a photo of the 1954 premiere of Eurovision, a Finnish board game about rail networks, sewing machines that could be purchased on installment (called the "American system"), an advertising postcard promoting Liebig meat extract ("Is it still meat?"). Each object-centered story is around 200 words in length and can stand alone. The stories are a pleasure to read. Badenoch and his co-curator-authors Suzanne Lommers (Lommers has managed the project since September 2012) and Sławomir Łotysz beautifully distill complex material for general audiences with wit and nuanced understandings of the objects' historical significance. Taken together, the composite view is one of a rapidly transforming and increasingly connected Europe, often showing how technology both brings people together and keeps them apart.

Users are invited to take one of 43 "tours," curated collections of stories on topics like "Expatriate Experts" and "Iron Curtain Computers." Thirty-five tours are written by the team, plus eight by guest curators affiliated with the network. Badenoch explains, "we kept the term 'tour' because of its implications of a selective and incomplete trajectory."[4] This fits: tours feel a bit like eclectic show-and-tell sessions. The tour "Fridges Change Kitchens," for example, draws on a photographed object, two historic photographs, two historic film clips, and a printed advertisement. Through these objects, the tour starts with the refrigerator's early industrial uses and goes on to show how it not only transformed households but was also used to represent Western technological development and affluence during the Cold War and to promote atomic power as promising abundant energy for peaceful purposes (Figures 2 and 3).

Although the linear tour structure makes navigation easy for users, it also underexploits the medium's potential for extensive internal (and external) linking. Of course, the Europeana API feeds an attractive composite of thumbnail images to the bottom of each page, inviting users to browse the collections of participating cultural heritage institutions. In addition to this, the tours could more frequently suggest detours to related digital resources outside of Europeana, such as the link to a retrodigitized copy of the atlas *Gesellschaft und Wirtschaft* in the "Knowing Everything" tour or the externally linked Chernobyl radiation diffusion map in the "Weather Mapping" tour. Such powerful historic or born-digital visualization tools enhance the site's content, even if they are hosted elsewhere.

The tour concept is familiar to users and creates opportunities for external contributions. User-generated content (tours, stories, or crowdsourced tagging) may be the most powerful way to stimulate public engagement in the project and include more detailed local, even personal, perspectives in the future. This model of allowing users to create collections and add stories in association with objects has already succeeded with global digital projects such as Historypin or digital

2.

Screenshot from *Inventing Europe: European Digital Museum for Science and Technology* (http://www.inventingeurope.eu), *Daily Lives* exhibition, tour "Fridge Changes Kitchens," story "The Birth of the Cool." Curator Alexander Badenoch. The displayed object is an Etna refrigerator (1935) that is in the collection of the Science Center NEMO in Amsterdam. Photo: Foundation for the History of Technology, Eindhoven (the Netherlands). Accessed 22 August 2014.

augmentations to brick-and-mortar museums such as the Regional Archive Leiden.[5] *Inventing Europe* plans to add user-contributed tours within the context of educational partnerships.

Collected from a consortium of cooperating museums, more than 1,000 objects ground these global narratives. The Eindhoven-based team has coordinated images of objects from ten European cultural heritage institutions, each with different missions and infrastructures: Science Museum (United Kingdom), Deutsches Museum (Germany), Norwegian Museum of Science and Technology (Norway), Museum Centre Vapriikki (Finland), Netherlands Institute for Sound and Vision (Netherlands), Science Center NEMO (Netherlands), Hungarian Museum of Science, Technology and Transport (Hungary), Institute for Tropical Research (Portugal), Museum Boerhaave (Netherlands), and Dokumentationszentrum Alltagskultur der DDR (Germany), as well as many more thumbnails fed through the Europeana API. In a future release, some of these institutional partners will contribute guided tours as well, presenting their objects in a new context.

In showcasing diverse collections, the creators aspire to "show multiple frames of curation" and emphasize objects' openness to multiple interpretations.[6] They have a good basis: each item

3.
Screenshot from *Inventing Europe: European Digital Museum for Science and Technology* (http://www.inventingeurope.eu),
Daily Lives exhibition, tour "Fridges Change Kitchens," story "The Birth of the Cool." The displayed object is an Etna refrigerator (1935) that is in the collection of Science Center NEMO in Amsterdam. Users can click on "more about this object" to examine each object in a detail view. Photo: Foundation for the History of Technology, Eindhoven (the Netherlands). Accessed 22 August 2014.

includes a photo of the contributing institution, metadata (as available), and a link to the object's site in its home database (when one exists). Although the site offers only basic information for most objects, this is still remarkable considering the diversity of institutions and their digital practices. More information is available on the partners' page, but it is not highly visible alongside the objects. One cannot help but wish for more data about the objects and more insight into the ways they are presented in their home collections. In direct association with each story, the curator-authors already have personal profiles that cultivate interest and a feeling of personal engagement; they identify them as situated individuals. Why not include a richer portrait of the contributing institutions more directly associated with their objects? Participating museums could provide larger photos of their institutions' collections or teams, a short introduction describing its missions and offerings—even, perhaps, a brief comment about what that object means in its local (or national) context.

To convey the objects' significance and the project's goals to the user, good web and interaction design is key. *Inventing Europe* was originally built on a simple, cost-effective Wordpress magazine-style template. An earlier prototype exhibition called *Europe, Interrupted,* used a tube map design to index site content (a trace is still visible in *Inventing Europe*'s header). This design associated texts with objects, telling stories along themed "lines." It was clever and graphically exciting but also somewhat overwhelming in its complexity. The new design, organized around the tripartite narrative structure, is more intuitive. The driving role of narratives, however, means that the objects take a back seat. The object-images could be promoted even further: although attractive, the thumbnail images are often too small. Larger images—even in the objects' detail views—would enliven the objects and attest to their importance, as would more information on object provenance, materials, and condition. The curator-authors' photos are a nice size; the object-images should be considerably larger.

Browsing by object is possible but not easy. A small link near the top left ("1000+ objects") offers this compelling alternative way to browse content by a giant composite of images. More than 1,000 objects are viewable en masse, awaiting inclusion in a future tour. Although it looks great, such a feature risks long loading times over some connections, especially if the collection grows. In the long run, it may be worth providing a way for users to filter the images (e.g., by date or home institution) so that they do not need to load all at once. Of course, this would require more complete metadata. Maps could be a future way to promote objects while reinforcing the theme of connecting Europe. An index map could plot points relevant to the objects (as appropriate) or map their home institutions. Alternatively, custom illustrative maps showing objects' movement or diffusion could serve as orientation for future guided tours. A post on the *Inventing Europe* Facebook page mentions a Europeana map search and display feature in development—perhaps this will become an option at a later point.

Guiding users by locating objects on maps and exhibition floorplans is a function of the new free Inventing Europe Museum App for iPhone and iPad,[7] developed courtesy of Museum Boerhaave and with funds from Eindhoven City Council, Eindhoven University of Technology, and the Foundation for the History of Technology. The app provides a tour of how inventions built Europe, via the *Starting with the Philips* exhibition at Museum Boerhaave in Leiden, expanding the collection's visibility while enhancing the museum experience for on-site visitors. The app promises regular updates to include content from other European museums. As it grows to include these partners, it will surely enhance connections between them, encourage common digital standards, and possibly even shared, or interwoven, narratives.

Diffusion of Knowledge

Compared to brick-and-mortar museums, printed monographs, or subscription-based databases, openly accessible digital projects can attract broad interest. Although *Inventing Europe* addresses both scholars and the public, it has an academically informed understanding of itself. It presents a sophisticated, reflective underlying concept and engaging narratives distilled from broad historical knowledge. So far, it has been propagated mainly among academic audiences, through its research network and academic conferences. As one would expect for a site addressing historians, *Inventing Europe* has quality references and useful citation tips. This could be taken further: adding a reference tool like Zotero would let academics add references directly to their own collections and share them among groups of interested users. And although the project reads relatively well on mobile devices, future versions may want to more explicitly design for tools used by a growing segment of visitors.

If the target audience is to extend beyond academics, users need to stumble upon it; the site is in need of search engine optimization and mutual linking with related sites. A big effect on search engine returns seems to come through the project's profile on the Europeana site (as a case study for implementing its API); for a time it was the top-ranked Google return for a search for "Inventing Europe." Between the summers of 2013 and 2014, queries for "Inventing Europe" and "Europe and technology" went from yielding no results for *Inventing Europe* on Google's first results page to ranking it as the top search result. This will surely drive more traffic to the site. Another way to increase exposure is through social media. The site is already primed for this by including tools for sharing via Facebook, Google+, Twitter, and Pinterest. For a new, small academic project it has been remarkably effective in reaching its core user base. In June 2013, the site attracted 850 visits per month, simply via word of mouth.[8] For the participating cultural heritage institutions, this is surely seen as success: reaching an international audience, even if specialized, creates visibility for national collections among a group that is likely to use and appreciate them. It blows dust off these objects and pushes them onto a global stage.

Essential to a digital project's impact is its hackability, that is, openness to (legal) further use and adaptation. Of course, *Inventing Europe* itself draws on other cultural heritage resources to make something new. Key to making a digital project useful for extension, adaptation, or combination is robust metadata. *Inventing Europe*'s decision to use Dublin Core metadata, one international standard, makes it more easily interoperable with other projects and opens opportunity for future interoperability. It displays its metadata prominently on each object detail page, underscoring its creators' awareness of its importance and the value of its contents as cultural heritage objects. Too often, however, the metadata are minimal. This is undoubtedly an effect of collecting data from ten diverse cultural heritage institutions, some of which do not yet have digital databases. Overcoming this lack of data (such as object sizes or publication dates) would require the local partners and the *Inventing Europe* team to research, supply, and enter more of the missing information. If more place and time metadata were to become available, much could be done to filter and index the site's content (although deciding which dates or locations may be relevant for the site's purposes is no clear-cut task). Hopefully, for future versions, the partners may be encouraged to provide more metadata. Users might even be tapped for metadata that can be more reliably crowdsourced, such as tagging objects with keywords.

More explicit sharing of the context and decisions behind not only metadata choices but also the conception and construction of the project as a whole would help others understand how they might use or build on the project further. As the project itself is a prime example of the kind of technological object it concerns, the "About" page would be an excellent place to reflect on the digital museum as a collector and also generator of knowledge. As a digital museum, does the project employ particular methods? The project is conscious of museum trends and practices (e.g., the "Vienna Method" is profiled in a story) yet could make public some reflections on its own. Some of this material can be found on *Inventing Europe*'s Facebook site, although the site appears to address a small network of academics.

The site's great value to the public lies in its educational potential, which has only begun to be cultivated. A link in the header's top right corner offers educational resources, currently more general suggestions for how the project may be used. In addition to its use by students in seventeen courses at European universities,[9] the site promises to soon share students' experiences and course resources—hopefully including more detailed examples of assignments. In the meantime, users may find value in the course outlines related to the exhibitions' themes. Also excellent is the idea for teachers to guide students through the creation of tours and propose to publish the most outstanding ones. But because most people find technology intimidating, a more welcoming invitation and simpler instructions for creating a tour—even allowing direct input into the content management system—might grease the rails. There are many ways forward: why not more heavily emphasize the proposed collaborative tours or promote conversations among students across borders? Such user-involvement strategies are essential to the project's future; user engagement and visibility grow hand in hand.

Concluding Assessment

Inventing Europe is an attractive and thoughtful project that succeeds in telling, and encouraging critical reflection on, the role of technology in building Europe. It takes big steps toward connecting and promoting collections at ten European cultural heritage institutions. Among its core academic users, it also achieves visibility for these museums by putting selected objects on a global stage, in a global context. The project is now ripe for expansion and promotion. As it matures, users will benefit from more of what is already there: more objects, more stories, more curators, more information about participating institutions, larger images, more metadata, more links, more interactivity. Of particular value will be more tours, from authors of the *Making Europe* book series, of course, but also from students and other users around the world. *Inventing Europe* has proven that historians and cultural heritage institutions can collaboratively harness digital humanities practices to build an attractive new resource—even in the complex context of Europe. It is on its way to becoming a model for how museums can attract broad interest and build connections globally.

Notes

1. Alexander Badenoch, "Translating Objects, Transnationalizing Collections: Inventing Europe between Museums and Researchers," in *Migrating Heritage: Experiences of Cultural Networks and Cultural Dialogue in Europe*, ed. Perla Innocenti (Farnham, UK: Ashgate, 2014), pp. 39–52. *Inventing Europe: European Digital Museum for Science and Technology*, www.inventingeurope.eu (accessed 22 August 2014).

2. Todd Pressner, Jeffrey Schnapp, Peter Lunenfeld, and contributors, "Digital Humanities Manifesto 2.0," 2009, http://www.humanitiesblast.com/manifesto/Manifesto_V2.pdf (accessed 16 July 2013).

3. Europeana is a virtual library presenting the scientific and cultural heritage of Europe from its pre- and early history to contemporary history by offering text, images, videos, and audio data. See the library's website, http://www.europeana.eu/ (accessed 24 August 2014).

4. Badenoch, "Translating Objects."

5. See Historypin, www.historypin.org (accessed 24 August 2014); and Regional Archive Leiden, www.archiefleiden.nl (accessed 24 August 2014).

6. Badenoch, "Translating Objects," referencing Mieke Bal, *Travelling Concepts in the Humanities: A Rough Guide* (Toronto: University of Toronto Press, 2002).

7. The Inventing Europe Museum App, https://itunes.apple.com/app/inventing-europe-museum-app/id828023607?mt=8 (accessed 24 August 2014).

8. Alexander Badenoch, e-mail message to author, 12 June 2013.

9. Suzanne Lommers (Inventing Europe project manager), e-mail message to author, 26 June 2013.

Bibliography

Badenoch, Alexander. "Translating Objects, Transnationalizing Collections: Inventing Europe between Museums and Researchers," in *Migrating Heritage: Experiences of Cultural Networks and Cultural Dialogue in Europe*, ed. Perla Innocenti (Farnham, UK: Ashgate, 2014), pp. 39–52.

Bal, Mieke. *Travelling Concepts in the Humanities: A Rough Guide.* Toronto: University of Toronto Press, 2002.

Pressner, Todd, Jeffrey Schnapp, Peter Lunenfeld, and contributors. Digital Humanities Manifesto 2.0. *Humanities Blast: Engaged Digital Humanities Scholarship*, 2009. http://www.humanitiesblast.com/manifesto/Manifesto_V2.pdf (accessed 16 July 2013).

About the Contributors

Johanna Conterio is a Postdoctoral Research Fellow at Birkbeck College, University of London, working with the Wellcome Trust project "The Reluctant Internationalists: A History of Public Health and International Organisations, Movements and Experts in Twentieth Century Europe." She studies environmental health and the environmental history of the Soviet Union, with a focus on the "subtropical" south and the Black Sea.

Kimberly Coulter directs the Environment & Society Portal, a digital platform for environmental humanities, at the Rachel Carson Center at the University of Munich. She earned a Ph.D. in geography at the University of Wisconsin–Madison and is interested in the transnational production and distribution of media and knowledge products. She coedits the blog *Ant Spider Bee*.

Bryan Dewalt is the Director of Curatorial Division and former Curator of Communications at the Canada Science and Technology Museums Corporation in Ottawa. He has published on the history of printing technology, photography, dictation machines, and digital networks.

Matthew Hockenberry is a media historian and technologist whose work examines the impact of media forms and material practice on production throughout the nineteenth and twentieth centuries. As a visiting scientist with the MIT Center for Civic Media, he developed Sourcemap, a platform for mapping supply chains, and he writes on the history of logistics and the world apparatus of production through its emblematic objects.

Harun Kaygan is an industrial designer by training and has a Ph.D. in architecture and design from the University of Brighton. His interests include design cultures, critique, and activism and, most recently, critical applications of new materialist theoretical frameworks to designed objects. He is currently Assistant Professor in Industrial Design at Middle East Technical University, Ankara, Turkey.

Rian Manson is a student at Carleton University, Ottawa, specializing in railway history and industrial technology. He spent the 2013–2014 academic year at the Universität Mannheim in Germany studying and researching East and West German steam locomotive design. He also interned in the Collections Department of the TECHNOSEUM in Mannheim.

David McGee received his Ph.D. in the history of science and technology from the University of Toronto and has worked on a variety of topics in the history of design ever since. He has been a Research Fellow at the Max Planck Institute for the History of Science in Berlin and a Fellow of the Dibner Institute for the History of Science and Technology at the Massachusetts Institute of Technology. He currently serves as Archivist of the Canada Science and Technology Museums Corporation.

Nina Möllers is currently leading a special exhibition project on the Anthropocene at the Deutsches Museum. Before that, she worked at the TECHNOSEUM in Mannheim and as a postdoctoral researcher at the Deutsches Museum on private energy consumption. She studied in Palo Alto, Tübingen, and Nashville and received her Ph.D. from the University of Trier in American history. Her research interests are the history of technology; environmental, gender, and museum studies; and the American South.

Allen Roda finished his Ph.D. at New York University in May 2013. His dissertation entitled "Resounding Objects: Musical Materialities and the Making of Banarasi Tablas" is available via his website, allenroda.com. He is currently a Jane and Morgan Whitney Research Fellow in the Department of Musical Instruments at the Metropolitan Museum of Art in New York.

Oliver Schmidt studied history, politics, and German at the universities of Münster and Hull. For his Ph.D., he researched the impact African American GIs had on Germany. After two years as Assistant Curator at the TECHNOSEUM in Mannheim, he is currently working on the development of an interactive exhibition on salt and its production and use in Westphalia, Germany.

Thomas Schuetz studied the history of science and technology in Stuttgart and Frankfurt. His research is focused on the knowledge transfer between the Islamic Orient and the Occident in premodern times and the history of environmental technologies and business in southwest Germany. He is a Research Associate at the newly founded Section for the History of the Impact of Technology (WGT) at the History Department of the University of Stuttgart, an endowed chair of the Berthold Leibinger Stiftung.

Knut Stegmann is working in the field of historical building research. From 2009 to 2013 he was a researcher at the Institute of Historic Building Research and Conservation of the ETH Zurich, Switzerland, and focused on the scientification of building knowledge in the nineteenth and early twentieth centuries. From 2007 to 2010 he worked on his doctoral dissertation at ETH Zurich dealing with the German building company Dyckerhoff & Widmann and its importance for the establishment of concrete constructions in Germany ("Das Bauunternehmen Dyckerhoff & Widmann—Zu den Anfängen des Betonbaus in Deutschland 1865–1918").

Index

Figures and captions are indicated by page numbers in *italics*.

01 14